Springer Oceanography

The Springer Oceanography series seeks to publish a broad portfolio of scientific books, aiming at researchers, students, and everyone interested in marine sciences. The series includes peer-reviewed monographs, edited volumes, textbooks, and conference proceedings. It covers the entire area of oceanography including, but not limited to, Coastal Sciences, Biological/Chemical/Geological/Physical Oceanography, Paleoceanography, and related subjects.

More information about this series at http://www.springer.com/series/10175

Alexander Zapevalov · Konstantin Pokazeev · Tatiana Chaplina

Simulation of the Sea Surface for Remote Sensing

Springer

Alexander Zapevalov
Marine Hydrophysical Institute
Russian Academy of Sciences
Sevastopol, Russia

Konstantin Pokazeev
Faculty of Physics
M. V. Lomonosov Moscow State University
Moscow, Russia

Tatiana Chaplina
Institute for Problems in Mechanics
Russian Academy of Sciences
Moscow, Russia

ISSN 2365-7677 ISSN 2365-7685 (electronic)
Springer Oceanography
ISBN 978-3-030-58751-2 ISBN 978-3-030-58752-9 (eBook)
https://doi.org/10.1007/978-3-030-58752-9

This Springer imprint is published by the registered company Springer Nature Switzerland AG
The registered company address is: Gewerbestrasse 11, 6330 Cham, Switzerland

Introduction

In the twenty-first century, the life of mankind is unthinkable without the widespread use of satellite methods, tools, and technologies both in the field of science and technology, and in everyday life. In the field of earth sciences, their use has made it possible to move from episodic observation of fragments of processes occurring in the World Ocean to studying it as a whole on a global scale in almost real time.

In the last two decades in the field of marine sciences and technologies, a new direction has emerged and is intensively developing, which is called operational oceanography. It is based on the synthesis of remote sensing and mathematical modeling together with the processing and presentation of observation data in real time in a user-friendly form. Operational oceanography relies on the extensive use of information obtained from satellite observing systems operating in the optical and radio ranges.

In remote sensing from spacecraft, information is "read" from the sea surface. The main provisions of the theory of scattering of electromagnetic waves at the ocean–atmosphere interface, as a component of the general theory of wave scattering on a rough surface, were formulated in the early 70s of the twentieth century. But the advances in scattering theory remain useless for ocean remote sensing tasks until reliable input data is obtained in the form of detailed and accurate statistical characteristics of the sea surface. The problem of wave scattering on a rough surface is one of the most difficult problems in mathematical physics. As applied to sea surface sensing, it becomes even more difficult because the sea surface is mobile.

Interpretation and possibilities of using remote sensing data are largely determined by models linking the topographic characteristics of the sea surface with processes in the boundary layers of the atmosphere and ocean. The field of short-period surface waves is very sensitive to such factors as wind, internal waves, current, stratification of the atmospheric boundary layer, upwelling, concentration of surfactants (both biogenic and technogenic), etc. Changes in the characteristics of the field of short-period surface waves create a fundamental possibility of remote monitoring of processes occurring in the atmosphere and ocean.

Correct interpretation of remote sensing data requires detailed information about the rough surface that forms the reflected signal. The first three chapters describe the statistical and spatial–temporal characteristics of the sea surface, focusing on the effects associated with the nonlinearity of sea surface waves. The analysis widely uses the data obtained by the authors on the stationary oceanographic platform of the Marine Hydrophysical Institute, installed in the Black Sea. The next seven chapters analyze how wave nonlinearity affects the formation of the reflected signal from the sea surface.

Chapter 1 analyzes the adequacy of the existing models of the probability density function of the sea surface elevations to the real wave field. The analysis is based on field measurements. The limitations and possibilities of using the probability density function of sea surface elevations, built on the basis of truncated Gram–Charlier series, are considered. The limitations are due to the fact that in real calculations, it is possible to use a relatively small number of members of the series, the coefficients of which are determined from the estimates of statistical moments. The use of truncated Gram–Charlier series results in negative values in probability density function models. A new approach is proposed that allows preserving the advantages of the Gram–Charlier distribution and reducing its disadvantages. It is noted that numerous attempts to construct the probability density function of sea surface elevations on the basis of hydrodynamic and kinematic models have not yet made it possible to adequately describe the wave field. This is due to the fact that there are a number of physical factors, such as inter-wave interaction, kinematic nonlinearity of waves of finite amplitude, mechanisms leading to the appearance of a group structure, and a number of others. When constructing hydrodynamic and kinematic models, as a rule, only one of these factors is taken into account.

In Chap. 2, sea surface slope variability is analyzed using 2D laser inclinometer measurements. The increased attention that is currently paid to the study of slopes is due to the fact that they are the main characteristic of the sea surface, which determines the processes of reflection and scattering of electromagnetic waves by the sea surface. The problem with studying slope variability is that the main contribution to their dispersion comes from short waves, whose characteristics are technically difficult to measure in the field. The most effective measurement tool that does not disturb the sea surface and has a high spatial resolution is a laser inclinometer. A series of experiments carried out on a stationary oceanographic platform yielded a number of interesting results.

The variability of slope characteristics in situations when slicks and ripples are simultaneously present on the sea surface are investigated. It is shown that in slicks the probability distribution is close to the Gaussian distribution, and in wind waves with increasing wind, the deviations from it increase. The process of changing the structure of the sea surface is not monotonic. The most abrupt changes in slope cumulants are observed in the transition from a smooth surface to a rough one; further changes are reduced mainly to a relatively slow increase in slope dispersion with an increase in wind speed. In artificial and natural slicks, the values of cumulants of the third and fourth orders of the two-dimensional distribution of slopes are close to each other and do not depend on the wind speed. The statistics of artificial and natural slicks differ from each other only in the dynamics of change in the variance of slopes, which is probably caused by differences in the physical processes that lead to their appearance.

In contrast to the distributions of the slope components, the distribution of the slope modulus in strongly differs from the model obtained under the assumption that the field of surface waves is linear. Regression dependences describing the relationship between the cumulants of the slope module and the wind were constructed based on the measurement data with a laser inclinometer.

Chapter 3 explores the spatial and temporal characteristics of sea waves. These studies are related to the recently emerged opportunity to obtain optical images of the sea surface with high spatial resolution and low time shift. The sequences of images allow calculating the phase velocities of surface waves, which are then used to reconstruct the current field and bathymetry of the coastal zone. When interpreting remote sensing data, it is necessary to take into account the nonlinearity of sea waves.

To study the spatio-temporal relationships, field experiments were carried out using arrays of string sensors of sea surface elevations. It was found that on the scales of dominant waves, the coherence in the longitudinal (relative to the general direction of wave propagation) direction decreases faster than in the transverse one. This effect cannot be explained in terms of a linear model.

The differences between laboratory and sea wind waves of the decimeter range are revealed. The difference lies in the fact that longitudinal coherence in laboratory waves is preserved at distances of up to 1–2 wavelengths, and in marine conditions, it disappears at much shorter distances (about half the wavelength). The effect of the rapid disappearance of coherence in the sea should lead to a weakening of the mechanisms of wave–wave interaction, which are realized due to the fulfillment of the synchronism condition.

Models are constructed that describe the phase relations and the quadratic coherence function in a multicomponent wave field.

Chapter 4 analyzes the shape of an altimeter radio pulse reflected from a quasi-Gaussian rough surface. The effect of changes in skewness and kurtosis of the rough sea level elevations observed in the field on the accuracy of sea level reconstruction is estimated. The deviations of the statistical moments of its elevations from the values corresponding to the Gaussian distribution resulting from the nonlinearity of sea surface waves lead to a change in the shape of the reflected altimeter pulse. Its leading edge is shifted, the position of which determines the travel time of the signal from the satellite to the sea surface and back, i.e., an error occurs in determining the sea level.

An important characteristic limiting the range of geophysical problems solved using altimetry data, along with their accuracy, is the spatial resolution. The use of synthetic aperture radars has significantly improved spatial resolution. In this situation, another physical factor must be taken into account. Remote sensing of sea surface level and significant wave height is performed under the assumption that surface elevation statistics are constant within the radar illuminated area. A fundamental property of sea waves is their group structure. If the dimensions of the irradiated area of the surface are comparable to or less than the wavelength of the wave group, the assumption of constancy of statistics ceases to be fulfilled, which leads to errors in reconstructing the surface level and significant wave height. The level determination error is proportional to the significant wave height. To ensure that the relative error in determining the significant wave height is no more than 10%, it is necessary that the following condition be met: the length of the section for which it is calculated should be more than twice the length of the wave group.

Chapter 5 discusses the features of Bragg (resonant) scattering of radio waves by the sea surface, caused by the presence of long surface waves. On the sea surface, there are always waves whose length is longer than the length of the Bragg waves. Therefore, short Bragg waves propagate along a curved surface, as a result of which the local angle of incidence changes. Situations when sounding is carried out in the centimeter and millimeter ranges are considered. The analysis of the effects created by long waves is carried out on the basis of data from direct field measurements of the sea surface slopes, carried out using a laser inclinometer. Quantitative estimates of the change in the backscattering cross section for sounding on horizontal and vertical polarization are obtained. The analysis took into account the deviations of the statistical distributions of the sea surface slopes from the Gaussian distribution. An important difference between the presented results is that they are obtained on the basis of calculations based on the data of direct field measurements of the sea surface slopes.

Chapter 6 discusses the possibility of determining the physicochemical characteristics of the sea surface from the data of active radio sounding in the microwave range. The main factor hindering the implementation of this approach is the dependence of the scattered radio signal on the sea surface roughness level varying over a wide range. Changes in the radio signal caused by changes in the spectrum of short waves are much greater than those caused by variations in temperature or salinity. To reduce the influence of this factor, it is proposed to use the polarization ratio. The sensitivity of the polarization ratio to changes in temperature and salinity is of the same order of magnitude as the sensitivity to these parameters of radiometric measurements. The advantage of passive sensing devices is their small size and lower cost. The advantage of active methods is the significantly higher spatial resolution.

Chapter 7 analyzes the accuracy of determining the statistical moments of sea surface slopes using optical scanners and lidars installed on spacecraft. Both methods are indirect, and the measurement accuracy is determined by the correctness of our ideas about the formation of an optical signal.

When using optical scanners, the calculation of the statistical moments of slopes is carried out by selecting seven parameters by minimizing the discrepancies between the empirical probability densities calculated from the images and the Cox–Munk model. The accuracy limitations are caused by two factors. The first factor is the measurement errors of the bidirectional reflection direction function in situations where sunlight is reflected toward the spacecraft by surface elements whose slopes have a low probability density. In this case, a significant contribution to the formation of optical images is made by physical mechanisms associated with the scattering of sunlight by the atmosphere, as well as with the penetration of sunlight under the atmosphere–ocean boundary and scattering in the water column. The condition that the optical image is formed only as a result of specular reflection is no longer met.

The second factor is related to the choice of the bidirectional reflection direction function, which includes the model of the probability density function of the sea surface slopes. For a nonlinear wave field, which is the field of sea waves, the slope probability density is approximated by a two-dimensional Cox–Munk model based on truncated Gram–Charlier series. The Cox–Munk model allows to correctly describe the slope probability density function in a limited range, beyond which the slope distribution distorts, negative values appear. Calculating the truncated distribution is a source of error.

In vertical laser sensing, slope dispersion calculations are performed under the assumption that the slope distribution is Gaussian and isotropic. Unaccounted for deviations of the slope distributions from the Gaussian distribution lead to a systematic error of 11–14%. The calculated values of the variance of the slopes turn out to be underestimated. The anisotropy of sea surface slopes, which cannot be measured with vertical sounding, leads to an error that also systematically underestimates the variance values. This error is on average less than 3%.

The influence of the duration of the probe pulse on the accuracy of determining the variance of the slopes is considered. It is shown that for the correct determination of the slope dispersion, it is necessary that the duration of the sounding pulses exceeds a certain characteristic time scale, which depends on the significant wave height. To measure the variance of slopes with a relative error of less than 5%, it is necessary that the duration of the probe pulse be twice the time scale equal to the time the laser pulse travels a distance equal to the significant wave height.

Chapter 8 analyzes the contribution to the variance of sea surface slopes from different wavelengths. Remote sensing data (radar, radiometric and optical measurements) as well as in situ measurement data (laser inclinometers, string sensors, wave floats) were used. An integrated approach made it possible to construct dependences, the dispersion of the sea surface slopes, created by waves in the range from the main energy-carrying waves to waves of a given length. It is shown that this dependence can be described by a power function. It is also shown that the contribution to slope dispersion from waves longer than 10 m is less than 20%.

Chapter 9 presents the results of a study of the variability of the ocean–atmosphere interface as a surface reflecting light. In the course of complex experiments, laser sounding of the sea surface from a stationary oceanographic platform was carried out simultaneously with measurements of its elevations and slopes and was accompanied by measurements of wind speed. It was found that the parameters characterizing specular reflection flares are stochastically related to the parameters of short gravitational and gravitational–capillary waves. Dependencies are constructed connecting the optical characteristics of the sea surface with its dynamic parameters and with the wind speed. Laser sensing makes it possible to reliably register structural anomalies appearing on the sea surface, created by an inhomogeneous and unsteady wind, sea currents, internal waves and Langmuir circulations, man-made pollution, and atmospheric precipitation.

Interesting results were obtained during the local upwelling period. In contrast to large-scale upwellings, with local upwelling, the level of sea surface roughness can decrease. This effect is the result of two oppositely directed processes affecting the roughness of the sea surface. When upwelling, cold, clear water rises to the surface. A decrease in the temperature of the surface layer of the sea leads to a decrease in the level of roughness, a decrease in the concentration of surfactants leads to its growth. In local upwelling, the second process may prevail.

Chapter 10 analyzes the physical limitations of the accuracy limit for surface wind speed measurements. The remote measurement of air flow parameters over the sea is based on the dependence of the sea surface roughness on the wind speed. Changes in roughness lead to changes in the power of the radio signal reflected from the sea surface, which are recorded on the satellite. The factor limiting the accuracy of remote determination of wind speed is the ambiguity of the relationship between the level of roughness and wind speed. In contrast to the standard approach, which compares the measurements of the wind speed from the spacecraft and from the buoy, here we investigate the direct relationship with the wind speed of the characteristics of sea surface waves that form the radio signal. It is shown that in the range of 4–15 m/s, it is possible to restore the wind speed with an accuracy of ± 1–2 m/s. If the wind does not exceed 4–5 m/s, with one-parameter wind recovery, the error becomes many times higher than with in situ anemometry. One of the reasons for this is the fluctuations in the levels of the spectral density of ripples with the appearance of burnishing zones (slicks). There is an additional fatal error, the existence of which is associated with the space–time inhomogeneities of the wind.

Some issues related to the influence of nonlinear effects in the field of sea surface waves on the results of remote sensing of the ocean in the radio and optical ranges were considered in a number of articles published in various journals. But to date, there is no monograph summarizing this problem. This book is geared for advanced level research in the general subject area of remote sensing and modeling as they apply to the coastal marine environment. It is of value to scientists and engineers involved in the development of methods and instruments of remote sensing, analysis, and interpretation of data. It will be useful for students who have decided to devote themselves to the study of the oceans.

Contents

Chapter 1
Statistical Distributions of Sea Surface Elevations

1.1 Introduction

In 1849, Stokes published a paper [28], which showed the kinematic nonlinearity of finite amplitude surface waves, but for a long time it was the linear model that remained the main model describing the field of sea surface waves. In the framework of the linear model, the surface wave field is represented as a sum of a large number of independent sinusoidal components, whose amplitudes are random variables, and the phases are randomly distributed with equal probability in the interval $(0, 2\pi)$. It follows from the linear model that both the surface elevation and its slopes are subject to Gauss distribution [21].

Active research into the non-linearity of the sea disturbance began in the early second half of the last century. It was shown theoretically that interaction between the components of the wave field leads to deviations in the distribution of its characteristics from the Gauss distribution [20, 25]. The work [17] seems to have shown for the first time experimentally that the distribution of the sea surface elevations is more consistent with the Gram-Charlier distribution than with the Gauss distribution. Field measurements confirmed deviations of statistical distributions of sea surface elevations from the Gauss distribution [2, 8].

Sea surface waves are slightly nonlinear, so deviations from Gauss distribution are small [9, 15, 34]. This allows classifying their distributions as quasi-gaussian distributions. Such distributions are described by probability density functions based on Gram-Charlier series [16], application of which has a number of features [32] that will be considered in this chapter.

Interest in the study of the statistical distribution of sea surface elevations is linked to the need to address a number of practical problems. Deviations from the Gaussian distribution influence the accuracy of sea surface level recovery when probing from spacecraft [10, 37], error of remote determination of significant wave height [3],

A. Zapevalov et al., *Simulation of the Sea Surface for Remote Sensing*, Springer Oceanography, https://doi.org/10.1007/978-3-030-58752-9_1

generation of infrasonic hydroacoustic radiation by sea surface [23, 35], generation of microseisms [7], probability of rogue waves [1, 11, 22] and others.

1.2 Possibilities and Limitations of the Gram-Charlier Distribution

The 2D problem of potential waves spreading along the surface of uncompressible inviscid liquid with the depth h is considered. In undisturbed state the liquid.

Currently, there are a large number of models built under various physical hypotheses and approaches that describe the probability density functions of sea surface elevations [5, 6, 12, 13]. But the principal method in the analysis of waveform measurements data [34] and in the description of the sea surface in the applications related to electromagnetic wave scattering is the distribution based on truncated Gram-Charlier series [4, 19, 26].

Let us consider the possibilities and limitations of using the Gram-Charlier distribution to describe the surface waves. It should be noted that the approach developed here will be fair for both the sea surface elevations and its slopes.

Let x—a random value. We introduce the notations μ_n is the moment of distribution of the order n; $P(x)$ is probability density function. Then

$$\mu_n = \int_{-\infty}^{\infty} x^n P(x) dx \tag{1.1}$$

If the mean value of a random value x is zero, then the first six cumulants of its distribution are connected with the moments of distribution by relations

$$\left.\begin{aligned} \lambda_1 &= 0 \\ \lambda_2 &= \mu_2 \\ \lambda_3 &= \mu_3 \\ \lambda_4 &= \mu_4 - 3\,\mu_2^2 \\ \lambda_5 &= \mu_5 - 10\,\mu_3\mu_2 \\ \lambda_6 &= \mu_6 - 15\,\mu_4\mu_2 - 10\,\mu_3^2 + 30\,\mu_2^3 \end{aligned}\right\} \tag{1.2}$$

Since the Gauss distribution moments are described by the ratios

$$\left.\begin{aligned} \mu_{2n} &= \tfrac{(2n)!}{2^n n!}\mu_2^n, \\ \mu_{2n+1} &= 0, \quad n \geq 1, \end{aligned}\right\} \tag{1.3}$$

then the condition is fulfilled for his cumulants

$$\left.\begin{array}{l}\lambda_n = \mu_n,\ n = 2 \\ \lambda_n = 0, \quad n > 2,\end{array}\right\} \tag{1.4}$$

Deviation of the values of senior random variables x ($n > 2$) from zero values is a sign of process non-linearity.

Gram-Charlier distribution based on a known decomposition in series by derived function [16],

$$PN(x) = \frac{1}{\sqrt{2\pi}} \exp\left(-\frac{1}{2}x^2\right) \tag{1.5}$$

Derived functions $PN(x)$ are defined by an expression

$$\frac{d^n}{dx^n} PN(x) = (-1)^n H_n(x) \cdot PN(x) \tag{1.6}$$

where $H_n(x)$ is Hermit's polynomials that have the property of orthogonality

$$\int_{-\infty}^{\infty} H_n(x) \cdot H_m(x) \cdot PN(x) dx = \begin{cases} 0 & npu\ n \neq m \\ n!, & npu\ n = m \end{cases} \tag{1.7}$$

With this approach, the density of the distribution can be represented

$$P_{G-C}(x) = PN(x) \cdot \sum_{m=0}^{\infty} \lambda_m H_m(x), \tag{1.8}$$

where λ_n is the coefficients determined from the observation data.

When modeling the probability density function of parameters (such as elevation, slope) of the sea surface, the Edgeworth shape of type A of the Gram-Charlier series [14] is used. The probability density function can be written in the following form

$$P_{GC}(x) = \frac{\exp\left(-\frac{x^2}{2\lambda_2}\right)}{\sqrt{2\pi\lambda_2}} \left[1 + \frac{\lambda_3 H_3\left(\frac{x}{\sqrt{\lambda_2}}\right)}{6\lambda_2^{3/2}} + \frac{\lambda_4 H_4\left(\frac{x}{\sqrt{\lambda_2}}\right)}{24\lambda_2^2} + \frac{\lambda_5 H_5\left(\frac{x}{\sqrt{\lambda_2}}\right)}{120\lambda_2^{5/2}} + \frac{\lambda_6 + 10\lambda_3^2}{720\lambda_2^3} H_6\left(\frac{x}{\sqrt{\lambda_2}}\right) + \cdots\right] \tag{1.9}$$

Cumulant λ_2 is a variance for a random value x. The first six polynomials of Hermit are described by the expressions

$$\left.\begin{aligned} H_0 &= 1 \\ H_1 &= x \\ H_2 &= x^2 - 1 \\ H_3 &= x^3 - 3x \\ H_4 &= x^4 - 6x^2 + 3 \\ H_5 &= x^5 - 10x^3 + 15x \\ H_6 &= x^6 - 15x^4 + 45x^2 - 15 \end{aligned}\right\} \quad (1.10)$$

In experiments, statistical moments not older than the fourth order are usually defined, so only the first five (including zero) members of the series are used when modeling the probability density function. For cumulant that satisfy the condition $n > 2$, we will introduce normalization $\tilde{\lambda}_n = \lambda_n/\lambda_2^{n/2}$. Then the model of the probability density function is written in the form of

$$\tilde{P}_{GC}(x) = \frac{\exp\left(-\frac{x^2}{2\lambda_2}\right)}{\sqrt{2\pi\,\lambda_2}}\left[1 + \frac{\tilde{\lambda}_3}{6}H_3\left(\frac{x}{\sqrt{\lambda_2}}\right) + \frac{\tilde{\lambda}_4}{24}H_4\left(\frac{x}{\sqrt{\lambda_2}}\right)\right] \quad (1.11)$$

where cumulant $\tilde{\lambda}_3$ and $\tilde{\lambda}_4$ are skewness and kurtosis. Thus, a truncated Gram-Charlier series is used to simulate the probability density function of sea surface elevations.

Enter the designation η is surface elevation. The main drawback of the model $\tilde{P}_{GC}(\eta)$ is that it can only be used in a limited range of sea surface elevations, when it η does not exceed some critical value η_{cr} which we will define below. A clear example of limitations in the use of the model (1.11) is, as shown in Fig. 1.1, the appearance of negative values of the function $\tilde{P}_{GC}(\eta)$ [34]. Distortions occur in the region of small values of the probability density function, but they can be of great importance in applications related, for example, to ocean remote sensing from spacecraft [26]. Also note that similar limitations in the use of models based on truncated Gram-Charlier series are observed when describing statistical distributions of sea surface slopes [29].

The boundary of the zone of occurrence of negative values of the function $\tilde{P}_{GC}(\eta)$ is the smallest by module root of the equation $\tilde{P}_{GC}(\eta) = 0$. We denote this root of the equation as η_b. The considerable dispersion of statistical moments of sea surface elevations observed in field studies determines the need for calculation the values η_b not by averaged but by actual values of cumulative $\tilde{\lambda}_3$ and $\tilde{\lambda}_4$. The results of calculations based on the measurement data obtained from the oceanographic platform [34] are shown in Fig. 1.2. We can see that negative values in approximation (1.11) at $\left|\eta/\sqrt{\lambda_2}\right| < 3$ appear regardless of the significant slope values $\varepsilon = \sqrt{\lambda_2}/L_0$. Situations when $\tilde{P}_{GC}(\tilde{\eta}) < 0$, at $\left|\eta/\sqrt{\lambda_2}\right| < 3$ are not observed.

The absence of negative values does not yet mean that the model $\tilde{P}_{GC}(\eta)$ corresponds to the field of distribution of sea surface wave elevations. Comparison of

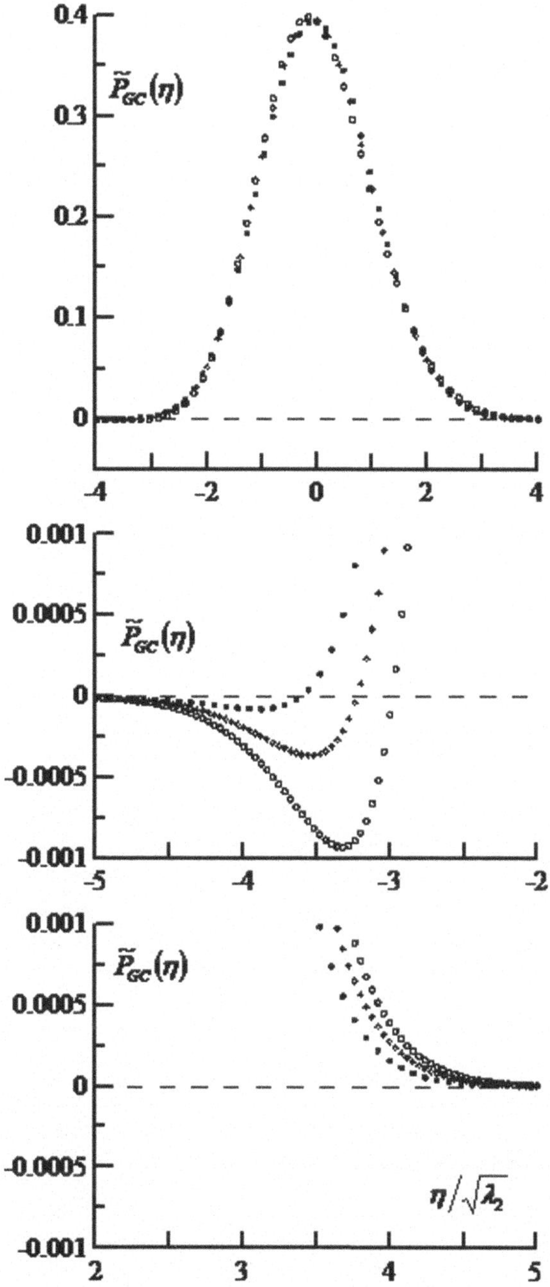

Fig. 1.1 Probability density function $\tilde{P}_{GC}(\eta)$

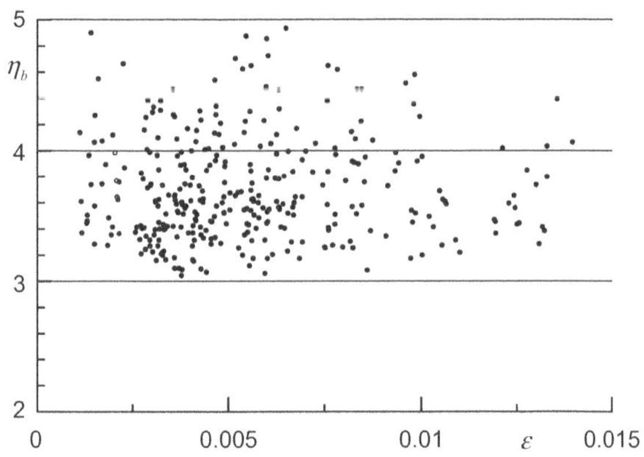

Fig. 1.2 Boundary of negative values appearance in the sea surface elevation probability density function $\tilde{P}_{GC}(\eta)$

the sea surface elevation histograms with the model calculations has shown that the model of the probability density function $\tilde{P}_{GC}(\eta)$ is only valid in the region where the condition is fulfilled [34].

$$\eta < \eta_{cr} \approx 2.5\sqrt{\lambda_2} \tag{1.12}$$

1.3 Combined Distribution Model

A new approach to the construction of the probability density function of quasi Gaussian processes in the field of surface waves, called the combined model, is proposed in the paper [33]. Originally, the combined model was intended for description of sea surface slopes. It is based on the synthesis of the Gauss distribution and the distribution built on the basis of truncated Gram-Charlier series. Within the range (1.12), the combined model corresponds to the Gram-Charlier distribution, while outside this range it approximates the Gaussian distribution. The general requirements that the model describing the distribution of sea surface elevations must satisfy are the following: single modality, the presence of no more than two inflection points, and the absence of negative values in the whole range of variation of random value [36].

The combined probability density function of the sea surface elevations can be presented as follows

$$P_C(\eta) = \frac{\exp\left(-\frac{\eta^2}{2\lambda_2}\right)}{\sqrt{2\pi\lambda_2}}\left\{1 + \left[\frac{\tilde{\lambda}_3}{6}H_3\left(\frac{\eta}{\sqrt{\lambda_2}}\right) + \frac{\tilde{\lambda}_4}{24}H_4\left(\frac{\eta}{\sqrt{\lambda_2}}\right)\right]F\left(\frac{\eta}{\sqrt{\lambda_2}}\right)\right\} \tag{1.13}$$

where the function $F\left(\eta/\sqrt{\lambda_2}\right)$ acts as a filter.

In the range (1.12), in which the model $\tilde{P}_{GC}(\eta)$ well describes the probability density function of elevations of the sea surface, the function $F(\eta)$ is close to one, at higher values $|\eta|$ it tends to zero. For compactness of mathematical expressions let us introduce the dimensionless parameter

$$\tilde{\eta} = \eta/\sqrt{\lambda_2}. \tag{1.14}$$

The two-parameter function has been selected as a filter F

$$F(\tilde{\eta}) = \exp\left[-(|\tilde{\eta}|/d)^n\right] \tag{1.15}$$

where parameter d defines an area within which $F(\tilde{\eta}) \approx 1$, the parameter defines the speed at which the function F tends to zero outside this area.

The combined model $P_C(\tilde{\eta})$ must satisfy the condition of "smoothness", i.e. the derivative $\frac{dP_C(\tilde{\eta})}{d\tilde{\eta}}$ must not change in a jumpy way. The nature of the function $P_C(\tilde{\eta})$ behavior in the vicinity of points $|\tilde{\eta}| = d$ determines the parameter n. As it grows n, the function $F(\tilde{\eta})$ approaches a rectangular window

$$F(\tilde{\eta}) = \begin{cases} 1 & if\ |\tilde{\eta}| \le d \\ 0, & if\ |\tilde{\eta}| > d \end{cases} \tag{1.16}$$

which causes the function $P_C(\tilde{\eta})$ to break at points $\tilde{\eta} = \pm d$. In the limit, the $n \to \infty$ transition from the Gram-Charlier distribution to the Gaussian is done in a jumpy way.

The density function of the probability of sea surface elevations reduction rapidly with growth $|\tilde{\eta}|$. To define the upper boundary of the range, in which the parameter n is set, let's consider the behavior of the function $f_C(\tilde{\eta}) = \frac{1}{P_C(\tilde{\eta})}\frac{d\,P_C(\tilde{\eta})}{d\tilde{\eta}}$. The type of the function $f_C(\tilde{\eta})$ is shown in Fig. 1.3. The diagonal coming from the upper right corner corresponds to the dependence $\frac{1}{P_G(\tilde{\eta})}\frac{d\,P_G(\tilde{\eta})}{d\tilde{\eta}} = -\tilde{\eta}$, where $P_G(x) = \frac{1}{\sqrt{2\pi}}\exp\left(-\frac{x^2}{2}\right)$—the Gauss distribution of a random value x, the dispersion of which is equal to one. With the growth of the parameter n, as follows from Fig. 1.3, local peaks appear in the vicinity of points $\tilde{\eta} = \pm d$ [33].

Select the condition of absence of local extrema in the function $f_C(\tilde{\eta})$ as the distribution smoothness criterion $P_C(\tilde{\eta})$. If changes in statistical characteristics occur within the limits corresponding to their standard deviations from the mean, the function $f_C(\tilde{\eta})$ has no local extrema when $n \le 3.5$.

The range of variation of n parameter are limited by two factors. The first factor: at small values of n there are deviations of distribution $P_C(\tilde{\eta})$ from the distribution $P_{GC}(\tilde{\eta})$ in the range (1.12), inside which the distribution $P_{GC}(\tilde{\eta})$ well describes the data of field measurements. The second factor: outside the range (1.12), negative values in the distribution $P_C(\tilde{\eta})$ may appear.

Fig. 1.3 The behavior of the function $f_C\left(\tilde{\xi}\right)$ at different values of the parameter n. curve 1—$n = 1$; curve 2—$n = 3.5$; curve 3—$n = 10$; curve 4—$-\tilde{\eta}$

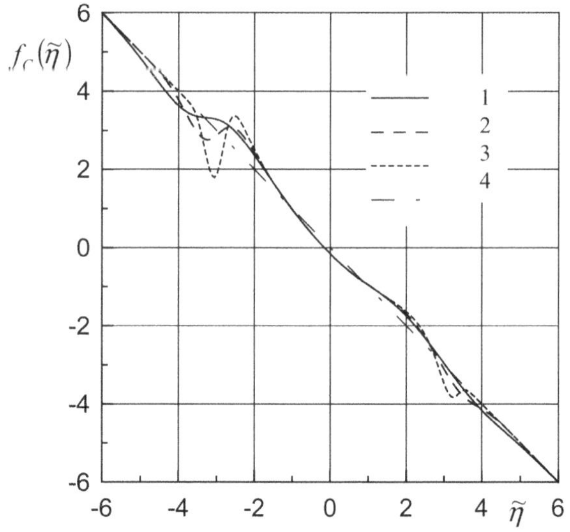

It should be noted that the combined model may also have negative values if the parameter d is incorrectly selected. An example of such situation is shown in Fig. 1.4, when negative values appear in the model $P_C(\tilde{\eta})$ if $d = 4.5$ and $d = 4$. When decreasing d, negative values do not appear in the combined model. In the

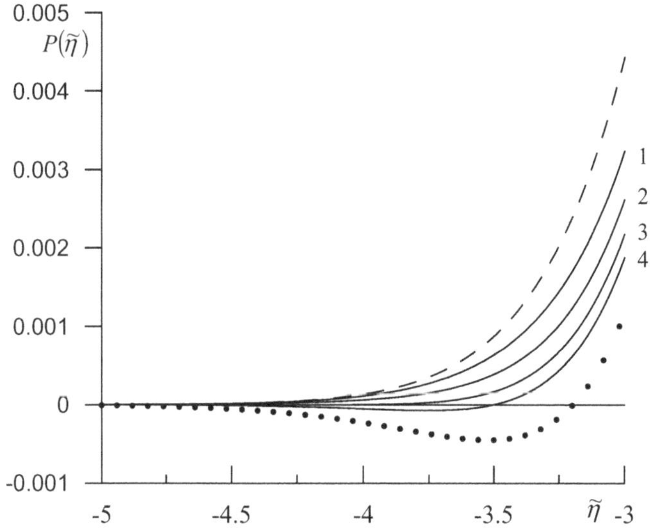

Fig. 1.4 Fragments of the probability density function of the elevation of the excited surface $P(\tilde{\eta})$: solid curves—combined model: 1—$d = 3$; 2—$d = 3.5$; 3—$d = 4$; 4—$d = 4.5$; dashed line is Gaussian distribution; dotted line is Gram-Charlier model, calculation at $\tilde{\lambda}_3 = 0.17$ and $\tilde{\lambda}_4 = -0.18$

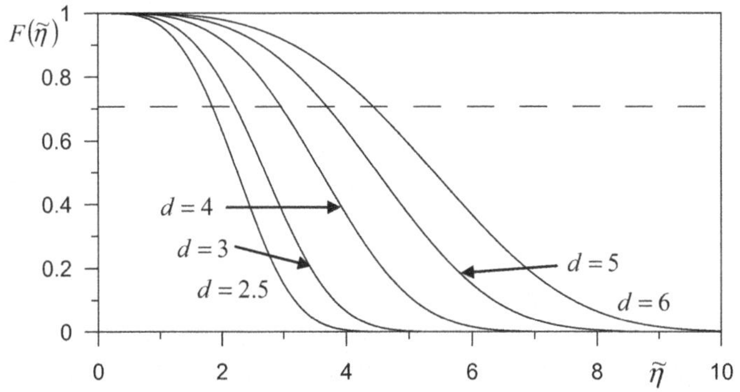

Fig. 1.5 Combined model filters $F(\tilde{\eta})$ calculated at $n = 3.5$. The dashed line shows the level $F(\tilde{\eta}_F) = 0.707$

range $|\tilde{\eta}| \geq 3$ the combined model is close to the Gaussian distribution and is positive everywhere.

We enter a parameter $\tilde{\eta}_F$ corresponding to the cutoff frequency in filters, which is determined by a condition accepted in radio engineering $F(\tilde{\eta} = \tilde{\eta}_F) = 0.707$. The filters $F(\tilde{\eta})$ constructed at $n = 3.5$ and different values of the parameter d are presented in Fig. 1.5. At the $n = 3.5$ the relation of parameters $\tilde{\eta}_F$ and d is described by the equation

$$\tilde{\eta}_F = 0.74\,d \tag{1.17}$$

The choice of parameter d is determined by the deviation of the simulated distribution from the Gaussian distribution. The larger the deviation is, the narrower the Gram-Charlier model is valid in the narrower range and, accordingly, the parameter d values should be lower. Since the model (1.11) based on a truncated Gram-Charlier series is fair within the range (1.12), the parameter value is d naturally assumed to be 3, which corresponds $\tilde{\eta}_F \approx 2.5$.

1.4 Sea Surface Cumulant Variability

The works [20, 25] have shown that deviations from Gaussian distribution caused by weak nonlinearity of sea surface waves depend on the significant slope of the sea surface. The significant slope can be defined as [14]

$$\varepsilon = \sqrt{\lambda_2}/L_0 \tag{1.18}$$

where L_0 is the dominant wavelengths. Sometimes instead of the significant slope, a similar parameter is used—steepness.

$$\upsilon = H_S/L_0 \tag{1.19}$$

where H_S is a significant wave height that equals the average height of the 1/3 highest waves [15]. The significant height of the waves is defined as $H_S = 4\sqrt{\lambda_2}$, respectively

$$\varepsilon = (1/4)\,\upsilon \tag{1.20}$$

Earlier it was shown that for the Gaussian distribution all odd statistical moments, third order and above, are equal to zero. The even statistical moments of the random n order with dispersion equal to one are equal $(2\,n)\,!/(2^n n\,!)$ at $n > 2$. The unambiguous values of statistical moments (or cumulants) of the Gauss distribution make them an effective tool for analyzing nonlinear processes in the wave field.

The conclusion based on mathematical modelling [20] that sea surface elevation cumulants depend on a parameter ε has been confirmed experimentally under laboratory conditions [14]. A series of laboratory experiments showed that cumulants up to and including order eight were indeed dependent on the significant slope. The exception was a fourth order cumulant (kurtosis), for which no dependence on the parameter ε was no revealed. Let us take a closer look at the changes in cumulants $\tilde{\lambda}_4$, $\tilde{\lambda}_5$ and $\tilde{\lambda}_6$.

According to the data from measurements in field conditions at low wind speeds and small fetch, the range of variation $\tilde{\lambda}_4$ is from -0.45 to $+0.45$ [18]. At significant wave height greater than 4.5 m, the values $\tilde{\lambda}_4$ are mostly in the range $-0.4 \leq \tilde{\lambda}_4 \leq 0.4$ [15].

Interest in studies of variability $\tilde{\lambda}_4$ has increased in recent years due to the fact that in experiments conducted in the Pacific Ocean, it was found that the ratio of maximum height of the waves to significant height increases with growth $\tilde{\lambda}_4$, i.e., the probability of abnormally large waves increases [22]. Anomalous waves have several names: freak waves, rogue waves, etc. Such waves include waves whose height is more than twice as high H_S [8]. Theoretical analysis of nonlinear dynamics of the field of knoidal waves also showed that the increase in the coefficient of $\tilde{\lambda}_4$ leads to an increase in the probability of abnormal waves [24].

According to measurements made on the oceanographic platform, with wind speeds ranging from 0.8 to 15 m/s, the values $\tilde{\lambda}_4$ are mostly in the range $-0.4 \leq \tilde{\lambda}_4 \leq 0.8$ [34]. The oceanography platform was installed in the Black Sea at a depth of 30 m. Measurements were made both under long fetch length when the wind blew from the open sea and in conditions of short fetch with coastal wind (length about 1 km). During the measurement period, the significant slope values ε varied within the limits of

$$0.0013 < \varepsilon < 0.018 \tag{1.21}$$

Wave records, randomly distributed over time, were carried out by sessions, the duration of which, usually, was 100 min. During statistical analysis, the wave records were broken down into 10-min fragments, each of which had its wind speed determined at the same time interval. For the same fragments the wave spectrum was calculated, by which the frequency of dominant waves ω_0 was determined. Then, using the dispersion equation for gravitational waves, the length of the dominant waves L_0 was determined and the significant slope was calculated.

Variability of cumulative values $\tilde{\lambda}_4$ determined from the measurement data on the oceanographic platform $\tilde{\lambda}_4$ when the significant slope of the sea surface ε changes is illustrated in Fig. 1.6. The lower boundary of the range in which this parameter $\tilde{\lambda}_4$ changes is close to the boundary of the range in which this parameter changed in earlier field experiments [15, 18], the upper boundary is markedly higher.

According to laboratory experiments, the values $\tilde{\lambda}_{4L}$ vary between −0.4 and −0.1 [14], i.e., they lie within the range in which the kurtosis changes under field conditions. Hereinafter, in order to distinguish the parameter estimates obtained in the field and laboratory experiments, the latter will be denoted by a lower index L.

Except for the work [32] we could not find works in which, for the field conditions, sea surface elevations cumulants of the fifth and sixth orders were defined. We will compare the results obtained in this work with data from laboratory experiments [14].

The dependence of the fifth order cumulant on the significant slope ε obtained under field conditions is shown in Fig. 1.7. The cumulant $\tilde{\lambda}_5$ is weakly correlated with the change in the significant slope ε. The correlation coefficient between the parameters $\tilde{\lambda}_5$ and ε is equal to 0.145. For the data array analyzed here, the half-width of the confidence interval for the zero correlation level at 97.5% is equal to 0.096. Thus, there is a weakly expressed growth $\tilde{\lambda}_5$ trend with increasing significant slope. This trend corresponds to a linear regression

$$\tilde{\lambda}_5 = 0.096 + 38.1\,\varepsilon \tag{1.22}$$

with a standard deviation of 0.51 [32].

The regression equation linking the parameters $\tilde{\lambda}_{5L}$ and ε which, according to laboratory experiments, looks like [14]

$$\tilde{\lambda}_{5L} = -110\,\varepsilon \tag{1.23}$$

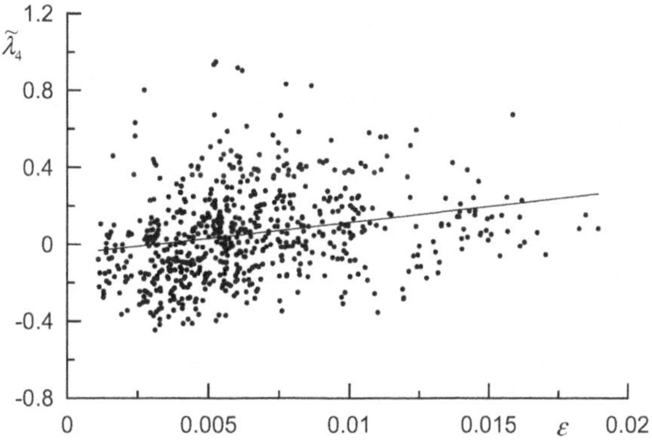

Fig. 1.6 Dependence of the cumulant $\tilde{\lambda}_4$ of sea surface elevation cult on the significant slope ε

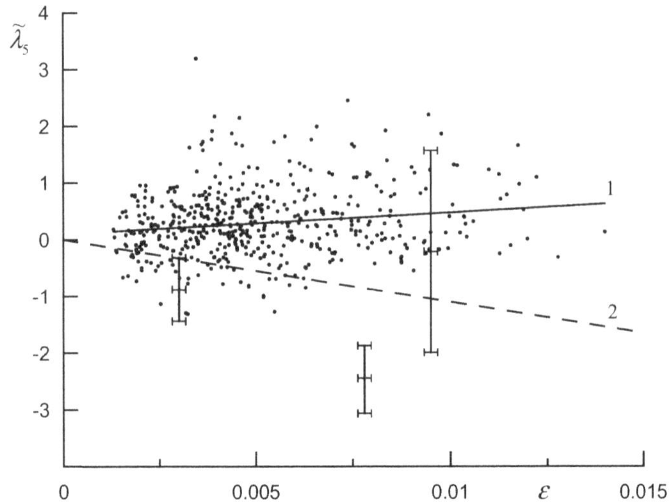

Fig. 1.7 The dependence of cumulant fifth order $\tilde{\lambda}_5$ on significant slope ε. Points—data of field measurements; curve 1—regression (1.22); curve 2—regression (1.23); vertical lines show values $\tilde{\lambda}_{5L}$ spread

is very different from Eq. (1.22). The coefficient at parameter ε in regression (1.23) is 2.9 times higher than in regression (1.22) and has the opposite sign. Probably, it is connected with that at carrying out of laboratory experiments the top limit of change of an significant slope reached values close to 0.5 that is three times higher than the maximum value of a slope received in field measurements. At the same time, significant cumulative changes $\tilde{\lambda}_{5L}$ were observed in the range of values ε greater than the maximum significant slope values obtained in field measurements (see Fig. 1.2) (see Fig. 1.7 [14]). In the range (1.21), in laboratory conditions, there was a large variation in the values of cumulants; in Fig. 1.8, this variation is shown by vertical lines.

The results obtained in field conditions show that changes in the sixth order of cumulative changes $\tilde{\lambda}_6$ with changes in the significant slope ε are uncorrellable. The value of the correlation coefficient between the parameters $\tilde{\lambda}_6$ and ε below the confidence value for zero correlation level with 97.5% confidence. In laboratory experiments, the regression equation linking the parameters $\tilde{\lambda}_{6L}$ and ε.

$$\tilde{\lambda}_{6L} = -5000\,\varepsilon^2 \tag{1.24}$$

The parameter changes are shown in Fig. 1.8 [32].

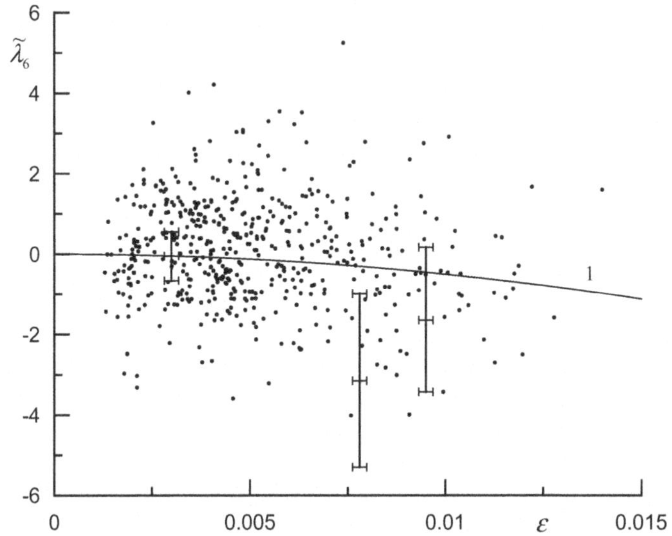

Fig. 1.8 Variability of the sixth order cumulant $\tilde{\lambda}_6$ and $\tilde{\lambda}_{6L}$. Points—field measurement data; curve 1—regression (1.24); vertical lines show the value $\tilde{\lambda}_{6L}$ spread

1.5 Testing of the Gram-Charlier Distribution

To determine whether models $\tilde{P}_{GC}(\tilde{\eta})$ can be used to describe the distribution of sea surface elevations, we compare it to the empirical probability density. As an empirical probability density, we will use a histogram normalized by the sampling length and the discharge width $P_E(\tilde{\eta})$. The discharge width is selected as 0.2 in the calculations $P_E(\tilde{\eta})$. The duration of the waveform was 10 min. To build the histograms, we used the same array of data that was used for the above mentioned analysis of simulants $\tilde{\lambda}_4$, $\tilde{\lambda}_5$ and $\tilde{\lambda}_6$.

Let us use the standard procedure of checking the hypotheses that the sample belongs to some distribution law [34]. We use a procedure based on the criterion of Kolmogorov's consent. As the null hypothesis, let us assume that

$$P_{EN}(\tilde{\eta}) = \frac{P_E(\tilde{\eta})}{1 + \frac{\tilde{\lambda}_3}{6} H_3(\tilde{\eta}) + \frac{\tilde{\lambda}_4}{24} H_4(\tilde{\eta})} = \frac{1}{\sqrt{2\pi}} \exp\left(-\frac{\tilde{\eta}^2}{2}\right) \tag{1.25}$$

i.e., we assume that the empirical distribution of the sea surface elevations transformed according to (1.25) corresponds to the Gaussian distribution. The correspondence between P_{EN} and P_G is estimated by the confidence probability

$$Pv\left(D\sqrt{N} > \delta\right) \approx P(\delta) = 2\sum_{n=1}^{\infty} (-1)^{n-1} \exp\left(-2n^2\delta^2\right) \tag{1.26}$$

where $D = \max|P_{EN}(\tilde{\eta}) - P_G(\tilde{\eta})|$; $\delta = D\sqrt{N}$—criterion of Kolmogorov's consent; N—number of measurements.

To build statistical estimates, independent observation (measurement) data are needed, so empirical distributions P_E were built on measurement data obtained with a discrete periodicity equal to or greater than half of the dominant wave period. The lengths of the dominant waves in the oceanographic platform area did not exceed 60 m during the research period. The frequency of the dominant waves f_0 during this period satisfied the ratio $f_0 \geq \min(f_0) = 0.16$ Hz. The discreteness of the wave measurements was chosen as $1/(2\min(f_0)) = 3.1$ s.

Calculation of Kolmogorov's consent criterion from an array of data, including 1850 ten-minute waveform measurement sessions, showed that $\delta = 5.7$. The corresponding value of the confidence probability $Pv \approx 0$. We conclude that the null hypothesis of correspondence of the sea surface elevation distribution to the model (1.11) does not agree with the measurement data and should be rejected.

Thus, the analysis shows that there are deviations in the empirical density of probabilities of sea surface elevations from the model (1.11), which should lead to errors in its use. Let us estimate these errors. For this purpose, let us analyze the relative error

$$\mathrm{O}(\tilde{\eta}) = \frac{P_E(\tilde{\eta}) - \tilde{P}_{G-C}(\tilde{\eta})}{\tilde{P}_{G-C}(\tilde{\eta})} \tag{1.27}$$

The function $\overline{\mathrm{O}(\tilde{\eta})}$ averaged by all measurements is shown in Fig. 1.9. When changing $\tilde{\eta}$ within the range $-3 < \tilde{\eta} < 3$, the values of function $\overline{\mathrm{O}(\tilde{\eta})}$ lie within the range from -0.02 to 0.07. Outside the specified range, the values of this function sharply increase [34].

Another characteristic determining the possibility of the model (1.11) for the analysis of the sea-wave field is the standard deviation of individual estimates of the function $\mathrm{O}(\tilde{\eta})$

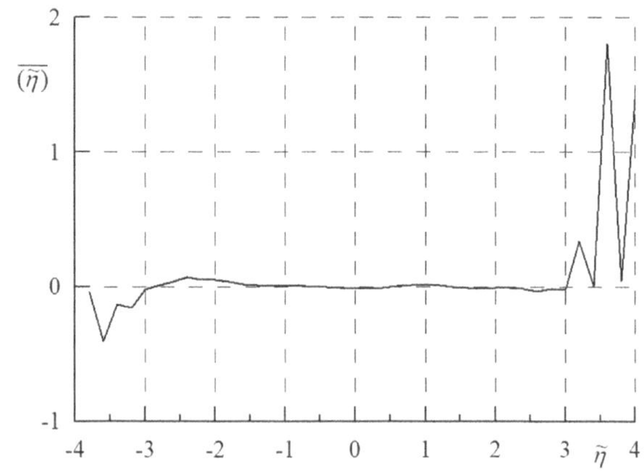

Fig. 1.9 Average value of relative error of the model functions of probability density $\overline{\mathrm{O}(\tilde{\eta})}$—$\Delta\,\mathrm{O}(\tilde{\eta})$—the standard value of relative error

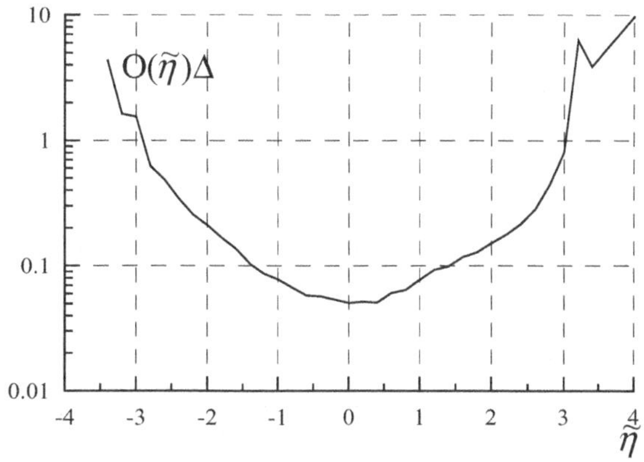

Fig. 1.10 RMS values of the relative error of the model probability density functions $\Delta\,\mathrm{O}(\tilde{\eta})$

$$\Delta\,\mathrm{O}(\tilde{\eta}) = \sqrt{\overline{\left(\mathrm{O}(\tilde{\eta}) - \overline{\mathrm{O}(\tilde{\eta})}\right)^2}} \tag{1.28}$$

As can be seen from Fig. 1.10, in the range $-1 \le \tilde{\eta} \le 1$ standard deviation is small, $\Delta\,\mathrm{O}(\tilde{\eta}) < 0.08$.

Condition $\overline{\mathrm{O}(\tilde{\eta})} < 0.3$ is met in the range $-2.5 < \tilde{\eta} < 2.5$. Outside the specified range $|\tilde{\eta}|$, the function $\overline{\mathrm{O}(\tilde{\eta})}$ grows rapidly, which determines the limitations of the model use (1.11).

1.6 Statistical Models of the Nonlinear Wave Field

There are a large number of models built within various physical hypotheses and approaches that describe the density function of probabilities of sea surface elevations [5, 6, 12, 13, 27, 30, 31, 36]. Let us consider some of them.

In recent years, an approach that combines statistical and dynamic methods has become common in the modelling of surface wave distribution [12]. Within the framework of this approach, a probability density function is constructed in the form

$$P_H(\tilde{\eta}) = \frac{1}{\sqrt{2\pi}}|1 - \upsilon\,\tilde{\eta}|\exp(-\upsilon\,\tilde{\eta})\,\exp\left(-\frac{1}{2}\,\tilde{\eta}^2\exp(-2\upsilon\,\tilde{\eta})\right) \tag{1.29}$$

where $\upsilon = \sqrt{\lambda_2}k$—dimensionless parameter associated with the significant slope of the ratio

$$\upsilon = 2\,\pi\,\varepsilon \tag{1.30}$$

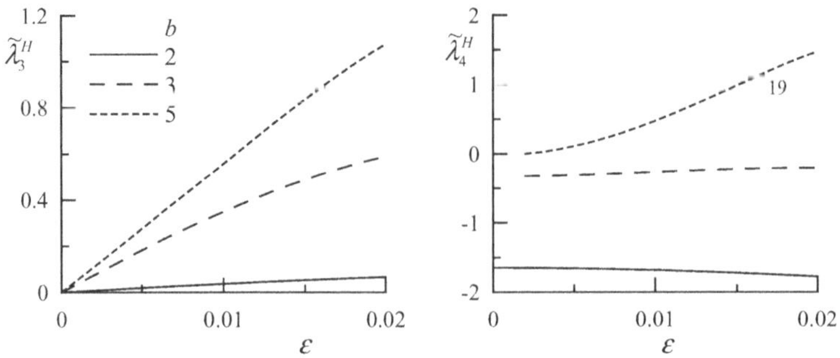

Fig. 1.11 Dependence of cumulants $\tilde{\lambda}_3^H$ and $\tilde{\lambda}_4^H$ the significant slope ε

At aspiration of a parameter υ to zero the model approaches to Gaussian distribution.

The principal disadvantage of the model $P_H(\tilde{\eta})$ is the following. The model $P_H(\tilde{\eta})$ is one-parameter. As a consequence, between the cumulants.

$$\tilde{\lambda}_3^H = \int_{-b}^{b} \tilde{\eta}^3 \, P_H(\tilde{\eta})\, d\tilde{\eta} \tag{1.31}$$

$$\tilde{\lambda}_4^H = \int_{-b}^{b} \tilde{\eta}^4 \, P_H(\tilde{\eta})\, d\tilde{\eta} - 3 \tag{1.32}$$

and the significant slope ε there is an unambiguous functional dependence, which as shown above is not observed in field experiments. Cumulants calculated with its help, as shown in Fig. 1.11, strongly depend on the limits of integration. With the growth of parameter *b*, they quickly go beyond the range determined in the experiments [15]. It should be noted that cumulants calculated for the distribution of Gram-Charlier, with the expansion of the integration limits are close to the values given in its construction.

Here for the convenience of comparison of the values of cumulants calculated by the model (1.29) with the data of field measurements in the construction of Fig. 1.11 instead of the parameter υ we used the significant slope ε.

The model $P_H(\tilde{\eta})$ built at different values of the significant slope is presented in Fig. 1.12. The left part of the figure clearly illustrates the behavior $P_H(\tilde{\eta})$ at low values $|\tilde{\eta}|$, the right part—at high values.

Another widely used method is based on the known model [28] describing the waveform of finite amplitude. The elevation of the sea surface at a point *x* created by each component of the wave field, can be recorded as a series

$$\eta(x,t) = a\,\cos X + \frac{a^2 k}{2}\cos 2X + \frac{3a^3 k^2}{2}\cos 3X + \cdots \tag{1.33}$$

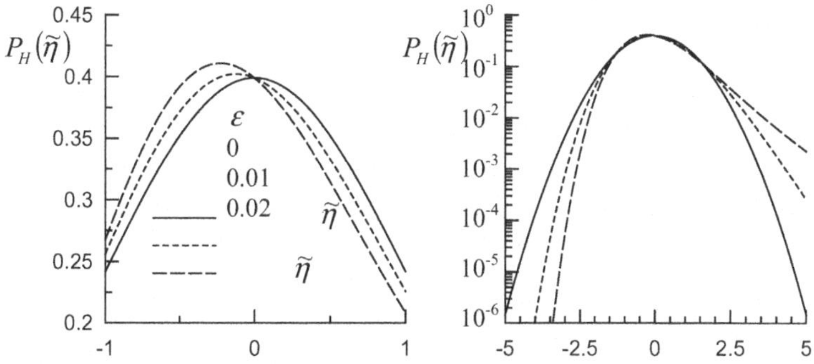

Fig. 1.12 Functions of the probability density of sea surface elevations $P_H(\tilde{\eta})$

where a is amplitude; t is time; $X = kx - \omega t + \phi$; k и ω is wave number and cyclic frequency; ϕ is phase. The amplitudes of the first order components are assumed to be distributed according to the Raleigh law, and the phases are evenly distributed.

The model of distribution of surface elevations for the Stokes wave in the third order approximation has the form [13].

$$P_S(\tilde{\eta}) = \frac{\exp(-h/2)}{\sqrt{2\pi}}\left[\frac{\vartheta}{\sqrt{R}} + \frac{9}{8}\frac{\lambda_2 k^2}{N}\frac{\vartheta}{R^{-5/2}} + \frac{5\,\lambda_2 k^2}{8\,NR^{-3/2}}\right] \tag{1.34}$$

where

$$N = 1 + \frac{1}{2}\lambda_2 k^2 \tag{1.35}$$

$$R = 1 + \sqrt{\lambda_2}\,k\;\tilde{\eta} + \frac{3}{4}\lambda_2 k^2 \tilde{\eta}^2 \tag{1.36}$$

$$\vartheta = N\left[1 - 2\sqrt{\lambda_2}\,k\;\tilde{\eta} + \frac{3}{8}\lambda_2\,k^2\,\tilde{\eta}^2\right] \tag{1.37}$$

$$h = N^2\left[\tilde{\eta} - \sqrt{\lambda_2}\,k\;\tilde{\eta}^3 + \frac{1}{2}\lambda_2\,k^2\,\tilde{\eta}^4\right] \tag{1.38}$$

Recently, within the framework of approximation of the second order of Stokes model, a simpler algebraic expression for the function of probability density of elevation has been proposed [31]. Without conducting a detailed analysis of the procedure for constructing these models, it should be noted that they, like the models $P_H(\tilde{\eta})$ [12], have fundamental shortcomings due to the fact that they are one-parameter. As a consequence, the third and fourth order cumulants calculated from these models are unambiguously linked, which contradicts the data of in situ measurements [15, 34].

1.7 Conclusion

The wide range of tasks related to ocean remote sensing requires detailed description of the characteristics of the sea surface. One of the characteristics of the sea surface that needs to be refined to improve the accuracy of remotely sensed parameters is the probability density function of elevations of the wave-covered sea surface.

At present, the main model in the applications related to the scattering of electromagnetic radiation at the ocean-atmosphere boundary remains the models of sea surface elevation distributions based on the truncated Gram-Charlier series. This model takes into account the weak nonlinearity of the sea surface wave field and describes its deviations from the Gaussian distribution.

The advantage of Gram-Charlier distributions is that in their construction the series coefficients are calculated on the basis of empirically determined statistical moments. Its application does not require additional conditions other than the standard conditions of statistical analysis, such as stationarity and homogeneity. The drawback of Gram-Charlier distributions is that when constructing them, information about statistical moments not older than the fourth order is usually available. Accordingly, a truncated Gram-Charlier series is used, which allows to describe a distribution of a random value only in a limited range of its changes. Outside this range, the probability density functions of the surface elevations are strongly distorted and may even have negative values.

There are numerous attempts to construct the probability density function of sea surface elevations based on hydrodynamic and kinematic models. However, they have not yet made it possible to adequately describe the wave field. This is due to the fact that there are a number of physical factors such as wave-wave interaction, kinematical nonlinearity of waves of finite amplitude, mechanisms leading to the appearance of a group structure and a number of others. In building such models, as a rule, only one of them, which determine the nonlinearity of sea surface waves, is taken into account.

This chapter deals with a new approach that preserves the merits of the distribution of Gram-Charlier and reduces its shortcomings. Initially, this approach was proposed to simulate sea surface slopes whose distributions belong to the quasi Gaussian class. The physical basis is that the nonlinearity of the sea surface field is weak, which in some cases allows to use a linear model of the wave field, which corresponds to the Gaussian distribution. A combined model is built that in the range of sea surface elevation, where the Gram-Charlier distribution is true corresponds to this distribution, and tends to the Gaussian distribution outside the specified range.

It should be added that although the approach to probability density function construction considered here was developed for the sea surface wave field, it can also be implemented for other weakly nonlinear processes.

References

1. Annenkov SY, Shrira VI (2013) Large-time evolution of statistical moments of wind-wave fields. J Fluid Mech 726:517–546. https://doi.org/10.1017/jfm.2013.243
2. Babanin AV, Polnikov VG (1994) On non-Gaussian wind waves. Phys Oceanogr 3:79–82
3. Basu S, Pandey PC (1991) Numerical experiment with modelled return echo of a satellite altimeter from a rough ocean surface and a simple iterative algorithm for the estimation of significant wave height. Proc Indian Acad Sci (Earth Planet Sci) 100(2):155–163
4. Callahan PS, Rodriguez E (2004) Retracking of Jason-1 data. Mar Geodesy 27:391–407. https://doi.org/10.1080/01490410490902098
5. Cieślikiewicz W (1989) Determination of the surface elevation probability distribution of wind waves using maximum entropy principle. Water wave kinematics. In: Proceeding of the NATO advanced research workshop on water wave kinematics. Kluwer, Molde, Norway, pp 345–352
6. Dai D, Wang W, Qian C, Sun F (2002) Comments on the surface elevation distributions derived by Huang. Appl Ocean Res 24(3):185–188
7. Dolgikh GI, Mukomel DV (2004) Dependence of microseism variation periods upon the cyclone propagation velocity and direction. Dokl Earth Sci 394(1):141–144
8. Fedele F (2015) On the kurtosis of deep-water gravity waves. J Fluid Mech 782:25–36
9. Guedes Soares C, Cherneva Z, Antão EM (2003) Characteristics of abnormal waves in North Sea storm sea states. Appl Ocean Res 25:337–344
10. Hausman J, Lotnicki V (2010) Sea state bias in radar altimetry revisited. Mar Geodesy 33(S1):336–347. https://doi.org/10.1080/01490419.2010.487804
11. Hjelmervik KB, Trulsen K (2009) Freak wave statistics on collinear currents. J Fluid Mech 637:267–284. https://doi.org/10.1017/s0022112009990607
12. Hou Y, Song G, Zhao X, Song J, Zheng Q (2006) Statistical distribution of nonlinear random water wave surface elevation. Chin J Oceanol Limnol 24(1):1–5
13. Huang NE, Long SR, Tung CC, Yuan Y, Bliven LF (1983) A non-Gaussian joint statistical model for surface elevation of nonlinear random wave fields. J Geophys Res 88:7597–7606
14. Huang NE, Long SR (1980) An experimental investigation of the surface elevation probability distribution and statistics of wind-generated waves. J Fluid Mech 101(1):179–200
15. Jha AK, Winterstein SR (2000) Nonlinear random ocean waves: prediction and comparison with data. In: Proceedings of 19th international offshore mechanics and arctic engineering symposium, ASME, 2000. Paper No. ETCE/OMAE 2000-6125
16. Kendall MJ, Stewart A (1958) The advanced theory of statistics. Vol. I. Distribution theory. Butler & Tanner Ltd., London, 675 p
17. Kinsman B (1965) Wind waves; their generation and propagation on the ocean surface. Prentice Hall Inc., Englewood Cliffs, N.J., 661 p
18. Kinsman B (1960) Surface waves at short fetches and low wind speed a field study. Chesapeake Bay Institute, Johns Hopkins University Technical Report. No. 19, 586 p
19. Lebedev NE, Pustovoitenko VV, Pokazeev KV, Melnikova ON (2014) Simulation of the sea surface bidirectional reflectance distribution function. Sovremennye Problemlemy Distantsionnogo Zondirovaniya Zemli Kosmosa 11(3):310–320 (in Russian)
20. Longuet-Higgins MS (1963) The effect of non-linearities on statistical distribution in the theory of sea waves. J Fluid Mech 17(3):459–480
21. Longuet-Higgins MS (1957) The statistical analysis of a random moving surface. Philos Trans R Soc (A) 249:321–387
22. Mori N, Janssen P (2006) On kurtosis and occurrence probability of freak waves. J Phys Oceanogr 36:1471–1483
23. Naugolnikh KA, Rybak SA (2003) Sound generation due to the interaction of surface waves. Acoust Phys 49(1):88–91
24. Pelinovsky EN, Shurgalina EG (2016) Formation of freak waves in a soliton gas described by the modified Korteweg–de Vries equation. Dokl Phys 61(9):423–426
25. Phillips OM (1961) On the dynamics of unsteady gravity waves of finite amplitude. Part 2. J Fluid Mech 11:143–155

26. Pokazeev KV, Zapevalov AS, Pustovoytenko VV (2013) The simulation of a radar altimeter return waveform. Mosc Univ Phys Bull 68(5):420–425. https://doi.org/10.3103/S0027134913050135
27. Srokosz MA (1998) A new statistical distribution for the surface elevation of weakly nonlinear water waves. J Phys Oceanogr 28:149–155
28. Stokes GG (1849) On the theory of oscillatory waves. Trans Camb Philos Soc 8:197–229
29. Tatarskii VI (2003) Multi-Gaussian representation of the Cox-Munk distribution for slopes of wind-driven waves. J Atmos Oceanic Technol 20:1697–1705
30. Tayfun MA (1980) Narrow-band nonlinear sea waves. J Geophys Res 85(C3):1548–1552
31. Tayfun MA, Alkhalidi MA (2016) Distribution of surface elevations in nonlinear seas. In: Offshore technology conference Asia. https://doi.org/10.4043/26436-ms
32. Zapevalov AS (2011) High-order cumulants of sea surface elevations. Russ Meteorol Hydrol 36(9):624–629
33. Zapevalov AS, Pustovoitenko VV (2010) Modeling of the probability distribution function of sea surface slopes in problems of radio wave scattering. Radiophys Quantum Electron 53(2):100–110
34. Zapevalov AS, Bol'shakov AN, Smolov VE (2011) Simulating of the probability density of sea surface elevations using the Gram-Charlier series. Oceanology 51(3):406–413
35. Zapevalov AS, Pokazeev KV (2016) Modeling the spectrum of infrasonic hydroacoustic radiation generated by the sea surface under storm conditions. Acoust Phys 62(5):554–558
36. Zapevalov AS, Ratner Y (2003) Analytic model of the probability density of slopes of the sea surface. Phys Oceanogr 13(1):1–13
37. Zapevalov AS (2012) Effect of skewness and kurtosis of sea-surface elevations on the accuracy of altimetry surface level measurements. Izv Atmos Oceanic Phys 48(2):200–206

Chapter 2
Statistical Distributions Sea Surface Slope

2.1 Introduction

The problem of electromagnetic wave scattering on a rough surface is one of the most difficult in mathematical physics. In the annex to the ocean-atmosphere boundary wave scattering, this problem becomes even more difficult, since the sea surface is mobile and its statistical characteristics are still not sufficiently studied. If the surface is simple enough, it can be described using different models: sinusoidal, saw-like, hemispheric or cylinder irregularities, etc. When the irregularities are caused by natural causes (in particular sea surface), an approach that uses statistical description of both the surface itself and the reflected radiation in the form of random fields is necessary.

Remote sensing techniques for studying sea surface slope are now widely used. Sounding is carried out in both the optical band [6, 4, 7] and the radio band [21, 3, 10]. The measurement data obtained in this way are indirect. The physical mechanisms that form the reflected field depend on the ratio of the wavelengths of the surface and the probing electromagnetic wave [5, 30] (Plant 2002). The correctness of the results obtained from remote sensing depends on a priori models describing the structure of the sea surface [31]. In order to develop remote sensing tools, direct measurements of sea surface slope are necessary.

Optical methods using laser sensors are the most effective for direct measurements of slope [17, 24]. They do not disturb the surface wave field and allow to measure the slope on the scale of capillary waves, which is impossible with the use of meters in contact with the surface.

Chapter 2 describes the results of a series of marine experiments to study the variability in the statistical characteristics of sea surface slope using a two-dimensional laser clonometer.

A. Zapevalov et al., *Simulation of the Sea Surface for Remote Sensing*, Springer Oceanography,
https://doi.org/10.1007/978-3-030-58752-9_2

2.2 Two-Dimensional Laser Slopemeter

The principle of operation of the two-dimensional laser slopemeter is based on the measurement in two planes of the laser beam deflection angles from the vertical when passing from under water of the rough water-air interface [17].

The optical scheme of the laser slopemeter used in experiments on the oceanographic platform is shown in Fig. 2.1 [35, 15].

The radiation source is a laser (1) located in the underwater part of the instrument. The wavelength of the laser beam is 0.63 μm. The laser is installed in an airtight container. The laser beam, reflected from the mirror (2), passes through a protective glass (3) and, crossing the sea surface, hits a lens assembled from two identical lenses (4). The focal length of the lens (4) is 600 mm. The beam is projected onto a matte screen (5) and concentrated by the lens (6). The photoconverter is a television camera (8). The camera is equipped with an interference filter (7) at a wavelength of 0.63 μm.

The angle of deviation of the laser beam from the vertical is determined by the local inclination of the sea surface at the point where the beam crosses it. The area of the laser spot on the undisturbed sea surface is about 2 mm^2.

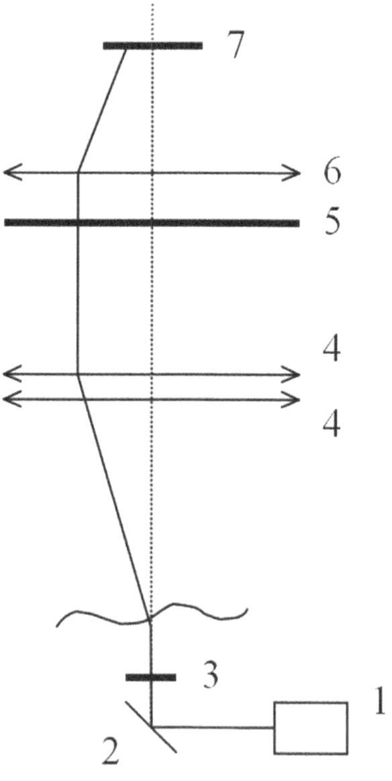

Fig. 2.1 Optical scheme of the laser slopemeter

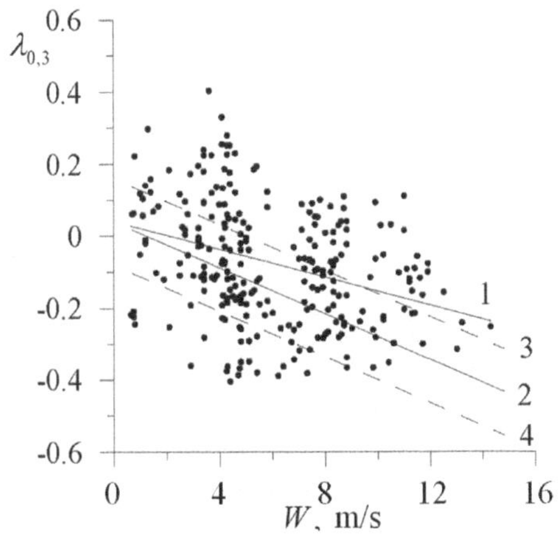

Fig. 2.2 Dependence of skewness of upwind component of slope $\lambda_{0,3}$ on wind speed W: points—experimental estimates, line 1 regression in the experiment [2]; lines 2, 3 and 4 regression and standard deviation from it in the experiment [35]

The advantage of the laser slopemeter over contact devices is that the laser beam provides measurement of the slope in a small area without disturbing the structure of the investigated wave field. It allows to measure the slope produced by capillary waves, which in marine conditions is not available to contact sensors. The wave crest collapse and foam appearance limit the possibilities to study the slope by laser slopemeters at high wind speeds.

Calibration of laser slopemeters is performed under laboratory conditions. Laser slopemeter used in experiments on oceanographic platform had the following main characteristics

- measured sea surface slope range $\pm 30°$;
- measurement error—0.2°;
- measurement discreteness—0.02 s;
- the range of permissible wave heights—up to 1 m.

There is a fundamental limitation in the use of laser slopemeters, due to the fact that the distance from the sea surface to the surface of the device can not be large. If this distance is large, the laser beam that has crossed the sea surface may not get into the lens of the photoreceiver.

2.3 Statistical Characteristics of Sea Surface Slope

2.3.1 *Distribution of a Weakly Non-linear Wave Field*

The slope of the sea surface are the first spatial derivatives of its elevations. When analyzing sea surface slope, the Cartesian coordinate system is usually used, with one of its axes oriented in the direction coinciding with the wind direction. Let us denote the elevation of the surface as $\eta(x, y, t)$, where x and y are coordinates in the horizontal plane. Let's assume that the slope components

$$\frac{\partial \eta}{\partial x} = \xi_u, \frac{\partial \eta}{\partial y} = \xi_c \tag{2.1}$$

are oriented respectively along and across the general wind direction (upwind and crosswind components). Hereinafter, for all slope characteristics, the index u corresponds to the upwind component and the index c to the crosswind component. If there is no lower index, the expression is true for both slope components. The components are ξ_u and ξ_c related to the slope module ξ_m (or the full slope) by relations:

$$\xi_u = \xi_m \cos\alpha; \xi_c = \xi_m \sin\alpha; \xi_m = \sqrt{\xi_u^2 + \xi_c^2} \tag{2.2}$$

where α—the azimuthal direction of slope.

When making in situ measurements with a two-dimensional slopemeter it is very difficult to install the instrument so that its orientation coincides with the general direction of wave propagation. To determine the upwind and crosswind components of the slope, the instrument rotates its coordinate system. The rotation of the coordinate system is performed by an angle at which the correlation coefficient between the orthogonal components of the slope is equal to zero. The dispersion ratio of the two orthogonal components reaches its maximum or minimum [8].

Field studies carried out using different types of equipment have shown that surface waves are a weakly nonlinear process, the distribution of their slope are quasi-gaussian [6, 21, 2, 14]. For approximation of the probability density function of such processes, models constructed with the use of Gram-Charlier series are usually used.

The Gram-Charlier rows are based on a known decomposition in series by derived function

$$PN(x) = \frac{1}{\sqrt{2\pi}} \exp\left(-\frac{1}{2}x^2\right) \tag{2.3}$$

The derivative functions $PN(x)$ are defined by the expression

$$\frac{\mathrm{d}^n}{\mathrm{d}x^n} PN(x) = (-1)^n H_n(x) \cdot PN(x) \tag{2.4}$$

where $H_n(x)$ is Hermit's polynomials are of the order n, which have the property of orthogonality:

$$\int_{-\infty}^{\infty} H_n(x) \cdot H_m(x) \cdot PN(x)\,\mathrm{d}x = \begin{cases} 0 & at\ n \neq m \\ n!, & at\ n = m \end{cases} \tag{2.5}$$

In a weakly nonlinear wave field, the probability density function of slope is described by the expression

$$P_{G-C}(\tilde{\xi}_c, \tilde{\xi}_u) = \frac{1}{2\pi}\exp\left[-\frac{1}{2}\left(\tilde{\xi}_c^2 + \tilde{\xi}_u^2\right)\right]\left\{\sum_{i,j} C_{i,j} H_i(\tilde{\xi}_c) H_j(\tilde{\xi}_u)\right\} \tag{2.6}$$

where $\tilde{\xi}_c = \xi_u/\sqrt{\overline{\xi_c^2}}$ и $\tilde{\xi}_u = \xi_u/\sqrt{\overline{\xi_u^2}}$ is normalised values of slope, $\overline{\xi_c^2}$ and $\overline{\xi_u^2}$ are the variance of crosswind and upwind components of slope. Thus positive ξ_u correspond to waves facing downwind, while negative ξ_u are waves facing upwind.

For certainty, we will consider that the first index of the coefficient $C_{i,j}$ corresponds to the slope in the crosswind direction, and the second—to the upwind direction. The coefficients $C_{i,j}$, where one of the indices is equal to zero, coincide with the coefficients of one-dimensional distribution of the corresponding slope component. The coefficients with $C_{i,j}$ both indexes not equal to zero are mixed. The order of mixed coefficients of the two-dimensional series is equal $i + j$.

2.4 Dependence of Slope Statistics on Wind Speed

The sea surface slope measurements analyzed in Chap. 2 were obtained from the stationary oceanographic platform of the Marine Hydrophysical Institute (Sevastopol). The platform has been located in the coastal zone of the Black Sea near the Southern coast of Crimea near the village of Katsiveli. The minimum distance from the platform to the shore is about 500 m. The depth at the place where the platform is installed is 30 m. The research results obtained on the oceanographic platform are compared with those obtained in different regions of the world ocean.

Estimates of the variance of slope components obtained in four independent experiments are given in Table 2.1. In the first experiment [2], the results of which have so far been widely used in remote sensing applications, the slope characteristics were determined at wind speeds of 0.7–14.8 m/s under open sea conditions. In the second experiment [6], statistical estimates of slope were obtained from the analysis of 24,000 images with the optical scanner POLDER (POLarization and Directionality of the Earth Reflectances) installed on the spacecraft. The third experiment [8] was conducted in the estuary of a river flowing into the Strait of Georgia. The slope were measured with a laser slopemeter aboard a slow moving vessel. A total

Table 2.1 Variance of sea surface slope components at different wind speeds

Parameter	Regression equation	Source
$\overline{\xi_c^2}$	$0.0026 + 0.00188\ W \pm 0.002$	Cox and Munk [2]
$\overline{\xi_u^2}$	$0.00034 + 0.00304\ W \pm 0.004$	
$\overline{\xi_c^2}$	$0.003 + 0.00185\ W \pm 0.0005$	Bréon and Henriot [6]
$\overline{\xi_u^2}$	$0.001 + 0.00316\ W \pm 0.0005$	
$\overline{\xi_c^2}$	$0.0008 + 0.00166\ W \pm 0.002$	Hughes et al. [8]
$\overline{\xi_u^2}$	$0.0015 + 0.00231\ W \pm 0.002$	
$\overline{\xi_c^2}$	$0.0014 + 0.00162\ W \pm 0.004$	Zapevalov [35]
$\overline{\xi_u^2}$	$0.0041 + 0.00205\ W \pm 0.005$	

of 9 series of measurements were made at wind speeds of 4.8–8.2 m/s. The fourth experiment was carried out on an oceanographic platform [35], which included 276 series of measurements with a laser slopemeter. The experiments were conducted at wind speeds of 0.6–14.3 m/s.

The regression equations obtained by the same methods are close to each other. At the same time, the variance sensitivity of the slope components to wind speed changes determined by a laser slopemeter is slightly lower than that obtained from optical images. The reason for lower values is as follows. Reflection of sunlight from the sea surface is determined by the slope created by all the waves present, while when measured by a laser slopemeter, the spatial resolution is limited by the size of the laser spot on the surface.

The variability of the third and fourth cumulants of the sea surface slope and their relation to the wind speed are shown in Table 2.2. Cumulants are marked with $\lambda_{i,j}$. The indexation introduced earlier for the coefficients $C_{i,j}$ in Eq. (2.6) is saved, the first index corresponds to the crosswind component of slope, the second index corresponds to the upwind component. The average value of the slope in any direction is zero. When calculating cumulants $\lambda_{i,j}$ in work [2] a priori, it was assumed that the distribution of the crosswind component of the slope is symmetrical, i.e.

$$\lambda_{3,0} = \lambda_{1,2} = \lambda_{1,3} = 0 \tag{2.7}$$

Note that the condition (2.7) is not always met with relatively short measurement sessions at a point or when restoring slope from the optical image of a small area of the surface. This occurs in a situation where the surface current is directed at an angle to the wind direction, or as a result of changes in the wind direction, since the timing of adjustment of the waves of different lengths to the changed meteorological conditions is different.

According to data [2, 8] the kurtosis of the upwind component of the slope does not depend on the wind speed and has an average value of 0.23 with a standard deviation of ±0.40. The estimates obtained on the oceanographic platform $\lambda_{0,4}$ mainly lie within

Table 2.2 Cumulants of the third and fourth order of slope of the sea surface

Parameter	Source	
	Zapevalov [35]	Cox and Munk [2], Hughes et al. [8]
$\lambda_{0,3}$	0.04 – 0.019 W ± 0.16	0.04 – 0.032 W ± 0.12 0.224 – 0.039 W ± 0.006
$\lambda_{1,2}$	−0.010 ± 0.12	0 0.016 ± 0.024
$\lambda_{2,1}$	0.0005 ± 0.09	0.01 – 0.0083 W ± 0.03 0.0086 ± 0.03
$\lambda_{3,0}$	0.036 ± 0.15	0 0.030 ± 0.05
$\lambda_{0,4}$	0.43 – 0.0369 W ± 0.12	0.23 ± 0.41 0.40 ± 0.15
$\lambda_{1,3}$	−0.012 ± 0.32	0 0.008 ± 0.03
$\lambda_{2,2}$	0.17 ± 0.27	0.12 ± 0.06 0.15 ± 0.08
$\lambda_{3,1}$	−0.05 ± 0.26	0 0.31 ± 0.059
$\lambda_{4,0}$	0.28 ± 0.29	0.40 ± 0.23 0.51 ± 0.29

the area of 0.23 ± 0.40, however, a weak dependence on wind speed is observed (Fig. 2.3). The correlation coefficient is equal to −0.38 with 95% confidence interval equal to 0.1. Relationship of the parameter $\lambda_{0,4}$ with wind speed is described by linear regression, shown in Table 2.2 (Fig. 2.2).

Values skewness $\lambda_{3,0}$ and kurtosis of $\lambda_{4,0}$ crosswind component of slope do not depend on wind speed (Figs. 2.4 and 2.5). They, as well as for cumulants λ_{03} and

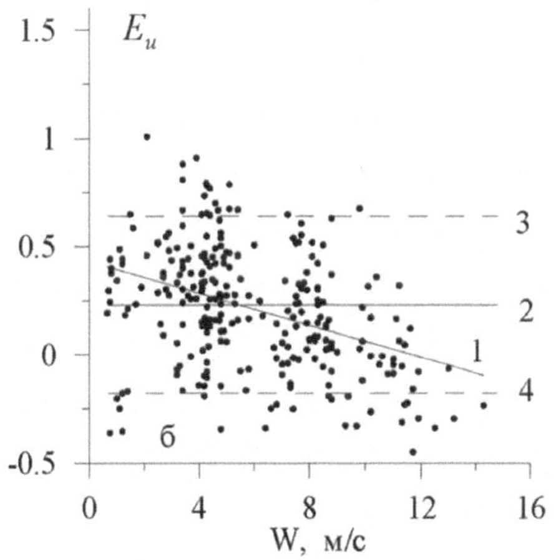

Fig. 2.3 Dependence of the kurtosis of the upwind component of slope $\lambda_{0,4}$ on wind speed W: points—experimental estimates, line 1 regression in the experiment [35]; lines 2, 3, 4 mean value and standard deviation in the experiment [2]

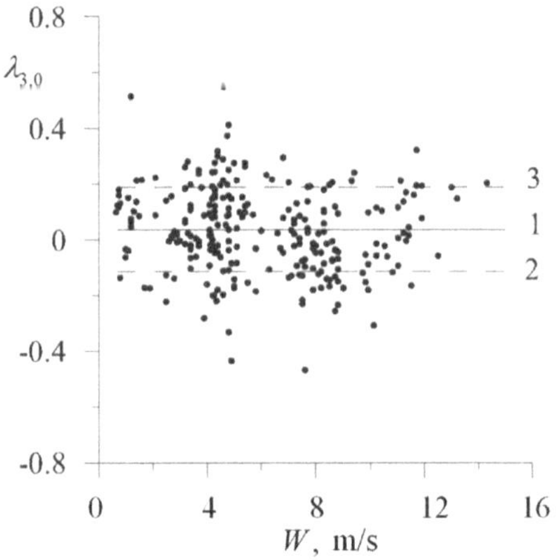

Fig. 2.4 Skewness of crosswind component of slope $\lambda_{3,0}$: points is experimental estimates, line 1 is average value, lines 2, 3 are standard deviations [35]

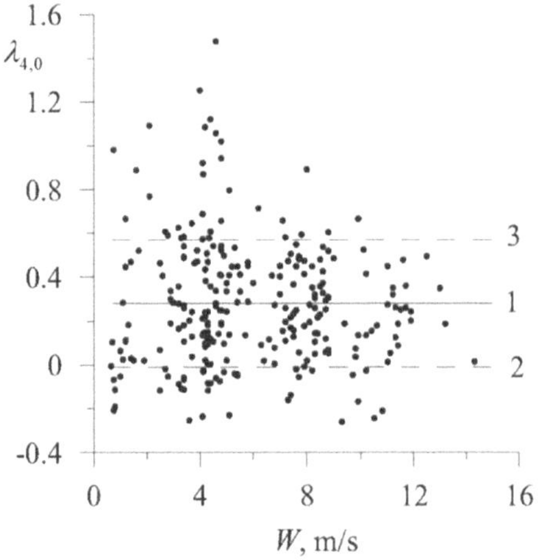

Fig. 2.5 Kurtosis of $\lambda_{4,0}$ crosswind component of slope: points are experimental estimates, lines 1, 2, 3 are mean value and standard deviations [35]

$\lambda_{0,4}$, are characterized by a significant variation. The skewness changes within the limits from−0.46 to 0.54, the kurtosis changes within the limits from −0.26 to 1.47 The average value is $\lambda_{3,0}$ close to zero and equals 0.036 at a standard deviation of 0.15. Average value of an kurtosis equals 0.28 at a standard deviation of 0.29.

Despite a good match between the average values of statistical characteristics, the variation of data in the fourth experiment is noticeably higher. This variance probably reflects the characteristics of coastal shelf dynamics and is caused by geophysical factors, in addition to common causes. These factors include the frequent presence of strong surface currents, the presence of several swell systems (most often two or three) in addition to wind waves, as well as features of the wind velocity field near the sea-land border.

The relatively small discrepancy between the results obtained from aerial photographs of the sea surface in the sun's glitter zone and those obtained from point measurements with a laser slopemeter suggests in favour of the hypothesis of the ergodicity nature of the sea wind wave. This means that it is possible to use the results of studies of the local characteristics of the slope to build models describing the formation of sea surface images in remote sensing.

2.5 Anisotropy of Sea Surface Slope

One of the most important characteristics determining the spatial structure of surface waves is the three-dimensional indicator introduced in Longuet-Higgins [19]. It is equal to the ratio of mean-square values of slope in crosswind and upwind directions.

$$\gamma = \sqrt{\overline{\xi_c^2}/\overline{\xi_u^2}} \tag{2.8}$$

For waves with long ridges the parameter γ is close to zero, for waves with short ridges it is close to one.

In practice, other characteristics similar to the three-dimensionality indicator, describing the anisotropy of slope, are applied. The works [2, 8] used a parameter $\gamma' = \overline{\xi_u^2}/\overline{\xi_c^2}$ associated with a γ simple algebraic ratio. The data presented in these papers indicate that the dependence of the parameter γ' on wind speed is weak. Similar results were obtained in experiments on an oceanographic platform. The values of the three-dimensional indicator determined in these experiments are shown in Fig. 2.6 [35].

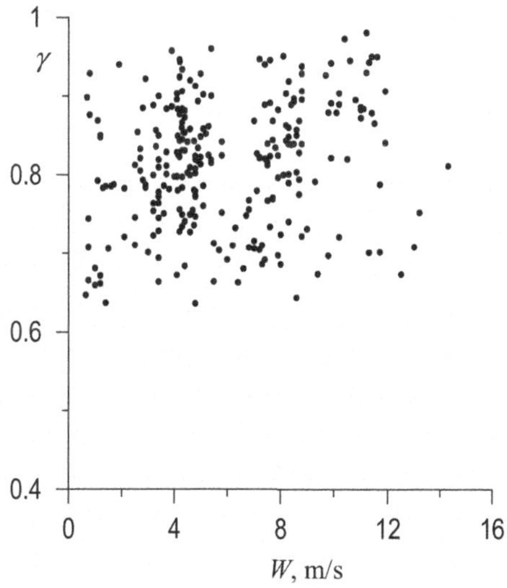

Fig. 2.6 Dependence of the three-dimensional indicator γ on wind speed W

The three-dimensional indicator γ is slightly dependent on wind speed. The correlation coefficient between parameters γ and W is 0.34. A positive correlation sign indicates that as the wind velocity increases, the values of the parameter γ on average increase slightly, i.e. the values of the standard-square values of the upwind and crosswind components of slope approach.

For the data on Fig. 2.6 of the three-dimensional indicator γ in the range of weather conditions from calm to wind speed 13 m/s its average value is 0.8. For comparison, we note that the values of the parameter γ' determined by aerial photographs [2] lie within the range from 1 to 1.8. The average value of the parameter γ' is 1.34, which corresponds to $\gamma = 0.86$.

2.6 Modeling the Probability Density Function of the Slope of the Sea Surface

2.6.1 Opportunities and Limitations of Distributions Based on Truncated Gram-Charles Series

To date, many models describing the scattering of acoustic and electromagnetic radiation from the sea surface have used its representation as a random moving Gaussian surface. A model for the mirror reflection points of a random moving Gaussian surface has been proposed in work [19]. Subsequently, the author himself noted the limitations of its application, in particular, this model does not allow to describe the sun's glitter shift in the presence of wind waves [20]. However, in the absence of detailed information on short surface waves in many practical applications, the ocean boundary is described as a random moving Gaussian surface [7].

Based on direct measurements of slope, it was shown that the sea surface is not strictly Gaussian. This is also confirmed by the results of remote sensing in the optical range [6, 36]. In the Gaussian model, it is also impossible to explain the inequality of backscatter cross-sections during radiosounding towards and in the wind direction [1].

Estimates of statistical sea surface slope moments from different principles indicate that their distribution belongs to the class of quasi-Gaussian distributions [6, 2, 14] (Su et al. 2002). When modeling the probability density function of similar processes, approximations based on the Gram-Charlier series are used. Let us consider how effective is the use of the Gram-Charlier series for describing the distributions of the slope components.

In the field, measurements of senior statistical moments of sea surface slope are associated with large errors. The Gram-Charlier series coefficients are calculated from empirical estimations of statistical moments (or cumulants), so only a truncated Gram-Charlier series can be used to simulate sea surface slope distributions. In experiments, statistical moments up to and including the fourth order are determined. Taking into account (2.7), instead of Eq. (2.6), for the probability density function of slope the equation in the form proposed in the paper [2] is used

$$P_{G-C}(\tilde{\xi}_c, \tilde{\xi}_u) = \frac{1}{2\pi} \exp\left[-\frac{1}{2}\left(\tilde{\xi}_c^2 + \tilde{\xi}_u^2\right)\right]$$
$$\left\{1 - \frac{1}{2}\lambda_{2,1} H_2\left(\tilde{\xi}_c\right) H_1\left(\tilde{\xi}_u\right) - \frac{1}{6}\lambda_{0,3} H_3\left(\tilde{\xi}_u\right) + \frac{1}{24}\lambda_{4,0} H_4\left(\tilde{\xi}_c\right)\right.$$
$$\left. + \frac{1}{24}\lambda_{0,4} H_4\left(\tilde{\xi}_u\right) + \frac{1}{4}\lambda_{2,2} H_2\left(\tilde{\xi}_c\right) H_2\left(\tilde{\xi}_u\right)\right\} \tag{2.9}$$

Approximation of the probability density function must meet several requirements. In particular, it must have such an important property as negativity. The model (2.9) has an imperfection because in certain range of slope it leads to negative probabilities (Tatarskii 2003). Let us define the boundaries of negative values in Eq. (2.9). For this purpose let's consider one-dimensional probability density functions of slope components

$$P_{G-C}^{(1c)}(\tilde{\xi}_c) = \frac{1}{2\pi} \exp\left[-\frac{1}{2}\left(\tilde{\xi}_c^2\right)\right]\left\{1 + \frac{1}{24}\lambda_{4,0} H_4\left(\tilde{\xi}_c\right)\right\} \tag{2.10}$$

$$P_{G-C}^{(1u)}(\tilde{\xi}_u) = \frac{1}{2\pi} \exp\left[-\frac{1}{2}\left(\tilde{\xi}_u^2\right)\right]\left\{1 - \frac{1}{6}\lambda_{0,3} H_3\left(\tilde{\xi}_u\right) + \frac{1}{24}\lambda_{4,0} H_4\left(\tilde{\xi}_u\right)\right\} \tag{2.11}$$

Two series of calculations were made. The first series used the average characteristics of skewness and kurtosis slope published in Cox and Munk [2]. Calculations showed that for the crosswind component of the slope, at $\lambda_{3,0} = 0$ and values $\lambda_{4,0}$ given in Table 2.2, the roots of the equation $P_{G-C}^{(1c)}(\tilde{\xi}_c) = 0$ are absent. However, negative values may exist for $P_{G-C}^{(1u)}(\tilde{\xi}_u)$. The type of functions $P_{G-C}^{(1u)}(\tilde{\xi}_u)$ at characteristic values of skewness and kurtosis of the upwind component is shown in Fig. 2.7. Negative values of the function $P_{G-C}^{(1u)}(\tilde{\xi}_u)$, appear in the range of large positive values $\tilde{\xi}_u$.

The skewness of the upwind component of slope depends on wind speed, so the values $\tilde{\xi}_u = \zeta_b$ at which the function $P_{G-C}^{(1u)}(\tilde{\xi}_u)$ becomes negative should also depend on W (Fig. 2.8).

In constructing Fig. 2.8, we used the linear regression dependence $\lambda_{3,0}$ on W proposed in the paper [2], this dependence is shown in Table 2.2. Within the framework of the assumption that the dependence $\lambda_{3,0} = \lambda_{3,0}(W)$ can be extrapolated to the region of higher wind speeds (measurements were made at $\leq$15 m/s), the analysis was extended to the wind speed range up to 20 m/s. To take into account the large variation $\lambda_{0,4}$ observed in the field experiments, the calculations were performed at several kurtosis values.

Considerable variation and lack of correlation between the cumulants $\lambda_{0,3}$ and $\lambda_{0,4}$ create the need to determine the values ζ_b not by averages but by actual values of skewness and kurtosis. In addition, as noted above, it is not uncommon to see situations when measured with a 2D laser slopemeter $\lambda_{3,0} \neq 0$. This leads to the

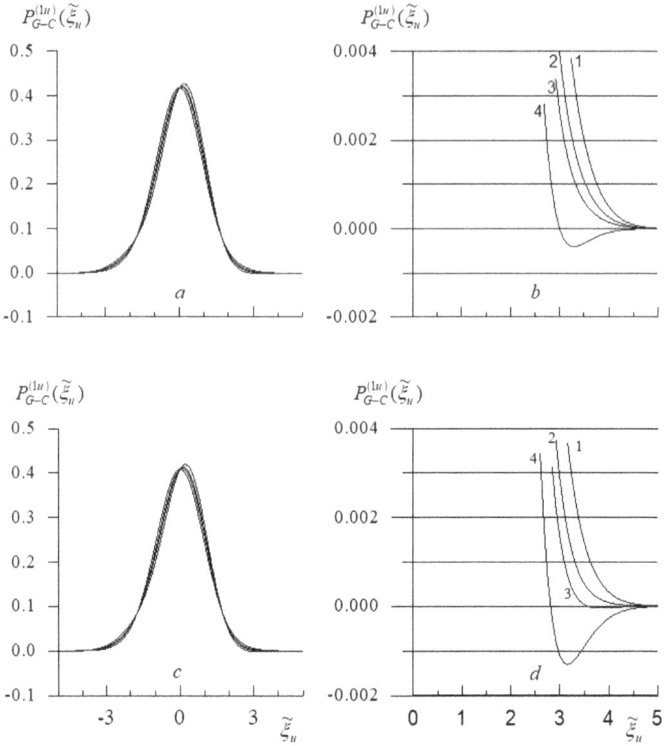

Fig. 2.7 Probability density functions of upwind component of slope in the form (2.11): **a**, **b**—$\lambda_{0,4} = 0.4$; **c**, **d**—$\lambda_{0,4} = 0.2$; curves 1–4 correspond to values $\lambda_{0,3}$ equal to 0, −0.2, −0.3, −0.5

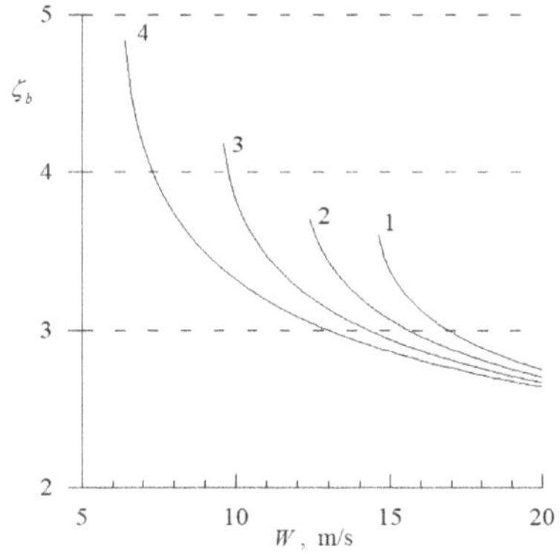

Fig. 2.8 Dependence of the boundary of negative values ζ_b appearance in approximation (2.11) on wind speed W: curves 1–4 are obtained with values $\lambda_{0,4}$ equal to 0.4, 0.3, 0.2, 0.1 respectively

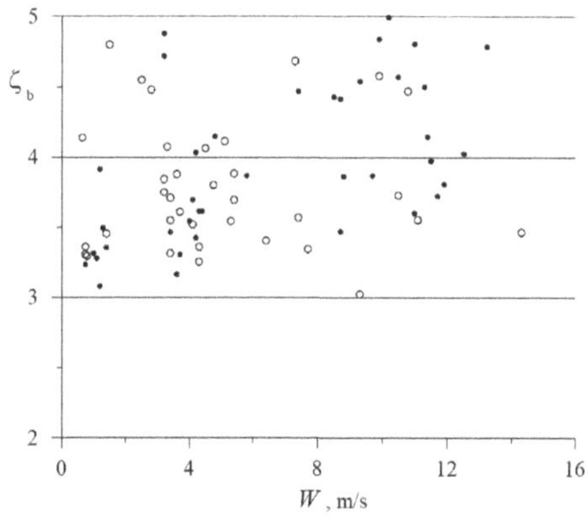

Fig. 2.9 Boundaries ζ_b of negative values appearance in the approximations of the probability density function of slope components: points are upwind component of slope, circles are crosswind

appearance of negative values in the approximation of the probability density function of the crosswind component of the slope. The results of calculations based on the actual values of cumulants are shown in Fig. 2.9.

An array of 276 measurement sessions was used to calculate the parameter ζ_b. In 25 cases for upwind component of slope and in 30 cases for crosswind value ζ_b were lower than 4. The appearance of values $\zeta_b < 4$ is observed at both high and low wind speeds. This applies to both the upwind and crosswind component of the slope. If $\tilde{\xi}_{u,} < 3$ and $\tilde{\xi}_{c,} < 3$ the slope probability density functions (2.10) and (2.11) are non-negative at all wind speeds.

It should be added that since the variance of slope increases with the wind speed, the slope range in which approximation (2.9) can be applied changes accordingly. For crosswind component of slope at wind speed of 5 m/s the application range of this approximation is ± 0.31, at 10 m/s is ± 0.45; for upwind component at the same wind speeds ± 0.40 and ± 0.52 respectively.

Although the application of the Gram-Charlier series to describe the distribution of sea surface slope is very attractive from a theoretical point of view, the practical application of approximation in the form of (2.9) is not possible in all hydrometeorological situations (Tatarskii 2003; Lebedev et al. 2014). The problem of building an analytical model of the probability density function of slope based on empirical estimates of the first four statistical moments remains open.

2.7 The Longuet-Higgins Model

The longwave components of the wind wave spectrum strongly influence the characteristics of short wind waves [23]. The paper [20] considers a number of factors affecting the distribution of sea wave slope. He showed that the main one is the amplitude modulation of short waves by a long wave, which may lead to a skewness of slope consistent in magnitude and sign with those observed in the experiments. The physical mechanism that leads to changes in the amplitudes of short waves along the longwave profile is the orbital velocity induced by it. The skewness of slope distribution is also caused by viscous attenuation of short waves, but the effect is small and can be neglected.

Let us represent the surface wave field as a superposition of short 2D waves and a one-dimensional long wave. We will consider that the long wave, which has an amplitude A, frequency ω_0 and wave number K, modulates the amplitudes of short waves with some phase shift ϕ_0. By setting the direction of the long wavelength propagation along the abscissa, the surface elevation can be represented as:

$$\eta(x, y, t) = A\cos(Kx - \omega_0 t) + \eta^{(s)}(x, y, t)\,[1 + M\cos(Kx - \omega_0 t - \phi_0)] \tag{2.12}$$

where $\eta^{(s)}(x, y, t)$—component of the surface elevation caused by short waves; M—modulation depth. The slope of the surface along the long wave direction are described by the expression:

$$\begin{aligned}\varsigma_x(x, y, t) = &-AK\sin(Kx - \omega_0 t) + \varsigma_x^{(s)}(x, y, t)\,[1 + M\cos(Kx - \omega_0 t - \phi_0)] \\ &- \eta^{(s)}(x, y, t)\,MK\sin(Kx - \omega_0 t - \phi_0)\end{aligned} \tag{2.13}$$

Let us accept the following assumptions: the amplitudes of short waves are much smaller than the amplitude of the dominant wave; the main contribution to the variance of slope give short waves; the modulation depth is less than one. Then the last term in (2.13) can be neglected, and we obtain:

$$\varsigma_x(x, y, t) = -AK\sin(Kx - \omega_0 t) + \varsigma_x^{(s)}(x, y, t)\,[1 + M\cos(Kx - \omega_0 t - \phi_0)] \tag{2.14}$$

The considered field of sea surface slope is a superposition of long wave and short wave slope. Short wave slope modulated by the long wave. Let us consider that in the absence of modulation the short wave slope are subject to the normal distribution law. In this case, the probability density function is described by the expression [33].

$$\begin{aligned}P_M(\varsigma_x) = \frac{1}{2\pi}\int_0^{2\pi} &\frac{1}{\sigma_x^{(s)}\sqrt{2\pi}[1 + M\cos(\phi - \phi_0)]} \\ &\exp\left\{-\frac{1}{2}\cdot\frac{[\varsigma_x + AK\sin(\phi)]^2}{\left[\sigma_x^{(s)}\right]^2[1 + M\cos(\phi - \phi_0)]^2}\right\}d\phi\end{aligned} \tag{2.15}$$

where $\left(\sigma_x^{(s)}\right)^2$ is the variance of the slope of short unmodulated waves; $\phi = Kx - \omega_0 t$ is the phase of the long wave. Note that the expression (2.15) can also be obtained in a more general case, when ϕ it is a random, slowly changing value, evenly distributed over the interval $[0, 2\pi]$. Thus, the considered model of sea surface slope leads to conditionally normal distribution.

Let us estimate how the modulation depth affects the variance σ_M^2, skewness A_M and kurtosis E_M of the distribution (2.15). To simplify the expressions, we introduce a dimensionless parameter $S = \frac{AK}{\sigma_x^{(s)}}$. After bulky tabs, which are omitted here, we get:

$$\sigma_M^2 = \left[\sigma_x^{(s)}\right]^2 \left(1 + \frac{M^2}{2} + \frac{S^2}{2}\right) \tag{2.16}$$

$$A_M = -\frac{3MS\sin\phi}{\sigma_m^3} \tag{2.17}$$

$$E_M = \frac{3\left[1 + 3M^2 + \frac{3M^4}{8} + S^2 + \frac{S^4}{8} + \frac{S^2M^2}{2}\left(\frac{1}{2} + \sin^2(\phi)\right)\right]}{\sigma_m^4} - 3 \tag{2.18}$$

The influence of the depth of modulation M on the values of parameters A_M and E_M shown in Fig. 2.10. The graphs are drawn for the values $S \leq 0.5$. The condition $S < 0.5$ corresponds to the fact that a greater contribution to the variance of slope give short waves.

Since the slope of the long wavelength in the crosswind direction is zero, it is necessary to accept in these expressions that $S = 0$. This, in particular, leads to a zero value of skewness of the crosswind component of the slope.

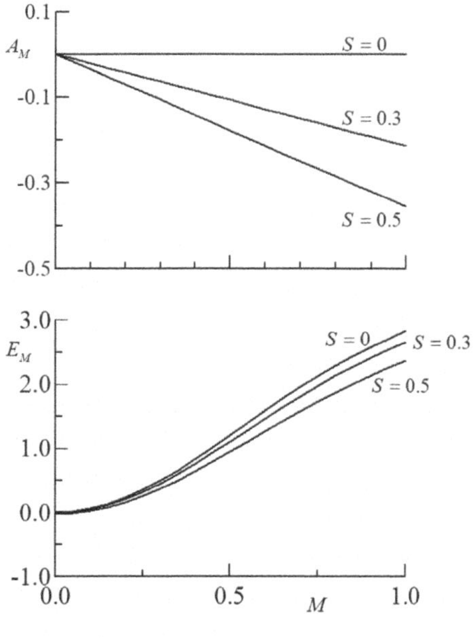

Fig. 2.10 Relations of skewness A_M and kurtosis of E_M slope distribution to the modulation depth M

The conclusion that can be drawn from the analysis is that modulation not only leads to skewness in the distribution of the upwind component of the slope, but also changes the fourth moment of both components.

2.8 Two-Component Conditional-Normal Distribution

In the paper [33] an analytical model of the probability density of sea surface slope was proposed, which allows to take into account the deviations observed in field experiments from the normal law of distribution. The model consists of two components with different weights,

$$P_{N-M}(x) = \varepsilon\, N(x, m_1, \sigma_1) + (1 - \varepsilon) \cdot N(x, m_2, \sigma_2) \tag{2.19}$$

where each component is described by an expression

$$N(x, m_j, \sigma_j) = \frac{1}{\sigma_j \sqrt{2\pi}} \exp\left[-\frac{(x - m_j)^2}{2\sigma_j^2} \right] \tag{2.20}$$

where m_j and σ_j are parameters of the j-th summand ($j = 1, 2$); ε is dimensionless parameter.

Model parameters (2.19) are calculated on the basis of a system of conditions to be met by its statistical momentum μ_i

$$\begin{cases} \mu_1 = 0 \\ \mu_2 = 1 \\ \mu_3 = A \\ \mu_4 = E + 3 \end{cases} \tag{2.21}$$

where A and E *are* set values for skewness and kurtosis. A special feature of the calculation is that there are five unknown parameters, we have only four conditions for moments. One of the parameters is free and should be set based on additional a priori considerations. As a free parameter, we use ε. Given the type of distribution (2.19), the values ε can be selected from a half-open interval [0.5, 1].

It follows from condition (2.21) that the statistical moments μ_i are both central and initial. The general expression for them is as follows:

$$\mu_i = \varepsilon\, \mu_{i,1} + (1 - \varepsilon)\mu_{i,2} \tag{2.22}$$

where

$$\mu_{i,j} = \int [x - \varepsilon m_1 - (1 - \varepsilon) m_2]^i N(x, m_j, \sigma_j)\, dx \tag{2.23}$$

the i-th moment for the j-th component.

Based on the known properties of normal distribution [11], after simple transformations we obtain the following analytical formula:

$$\mu_i = \sum_{k=0}^{[i/2]} (2 \cdot k - 1)\,!! \cdot C_i^{2k} \cdot (m_1 - m_2)^{i-2k} \left[\varepsilon (1-\varepsilon)^{i-2k} \sigma_1^{2k} + (-1)^{i-2k} \varepsilon^{i-2k} (1-\varepsilon) \sigma_2^{2k} \right] \tag{2.24}$$

Taking into account (2.22), to find the distribution parameters (2.19) it is necessary to solve a system of four algebraic equations:

$$\begin{cases} m \equiv \varepsilon\, m_1 + (1-\varepsilon) m_2 = 0 \\ \sigma^2 \equiv \varepsilon\,\sigma_1^2 + (1-\varepsilon)\sigma_2^2 + \varepsilon(1-\varepsilon)\,(m_1 - m_2)^2 = 1 \\ \mu_3 \equiv (m_1 - m_2)^3 \left[\varepsilon(1-\varepsilon)^3 - \varepsilon^3 (1-\varepsilon) \right] \\ +3(m_1 - m_2) \left[\varepsilon(1-\varepsilon)\sigma_1^2 - \varepsilon(1-\varepsilon)\sigma_2^2 \right] = A \\ \mu_4 \equiv (m_1 - m_2)^4 \left[\varepsilon(1-\varepsilon)^4 + \varepsilon^4 (1-\varepsilon) \right] + 6\,(m_1 - m_2) \\ \times \left[\varepsilon(1-\varepsilon)^2 \sigma_1^2 + \varepsilon^2 (1-\omega) \sigma_2^2 \right] + 3\left[\varepsilon\sigma_1^4 + (1-\varepsilon)\sigma_2^4 \right] = E + 3 \end{cases} \tag{2.25}$$

Let's consider two situations when $A \neq 0$ and $A = 0$. In the first situation, a sequential exclusion of parameters from the system of Eq. (2.25) can be obtained algebraic equation of 6-th degree relative to m_2

$$2\,\frac{\varepsilon^3 - 2\varepsilon^2 + 2\varepsilon - 1}{\varepsilon^3} m_2^6 + 4\frac{2\varepsilon - 1}{\varepsilon} A\, m_2^3 - 3E\, m_2^2 + \frac{\varepsilon}{1-\varepsilon} A^2 = 0 \tag{2.26}$$

Once defined m_2 other parameters can be found by the formulas

$$m_1 = -\frac{1-\varepsilon}{\varepsilon} m_2 \tag{2.27}$$

$$\sigma_1 = \sqrt{1 - \frac{1-\varepsilon^2}{3\,\varepsilon^2}\, m_2^2 - \frac{A}{3\, m_2}} \tag{2.28}$$

$$\sigma_2 = \sqrt{1 - \frac{2-\varepsilon}{3\,\varepsilon} m_2^2 + \frac{\varepsilon\, A}{3\,(1-\varepsilon)\, m_2}} \tag{2.29}$$

For the second situation ($A = 0$), corresponding to a crosswind slope component or a upwind component in weak winds, the solution is simpler. In this case $m_1 = m_2 = 0$, and from (2.25) we find

$$\sigma_1 = \sqrt{\frac{1}{\varepsilon}\left\{ 1 - (1-\varepsilon) \cdot \left[1 \pm \sqrt{\frac{\varepsilon}{3(1-\varepsilon)} E} \right] \right\}} \tag{2.30}$$

$$\sigma_2 = \sqrt{1 \pm \sqrt{\frac{\varepsilon}{3(1-\varepsilon)}}E} \tag{2.31}$$

The necessary conditions that the model (2.19) must meet are as follows: it must be non-negative, single-mode, and have no more than two inflection points [33]. To verify that the model (2.19) corresponds to the distribution of slope components in the real wave field, it has been compared with the empirical probability density function P_E obtained from the histogram of sea surface slope components. Comparison was carried out for both crosswind and upwind slope components. For the upwind slope component, two histograms corresponding to wind speeds of 3 and 10 m/s were constructed (Fig. 2.11). It is clearly seen that there is a deviation of the empirical distribution from the Gauss distribution. The deviation increases with the wind speed.

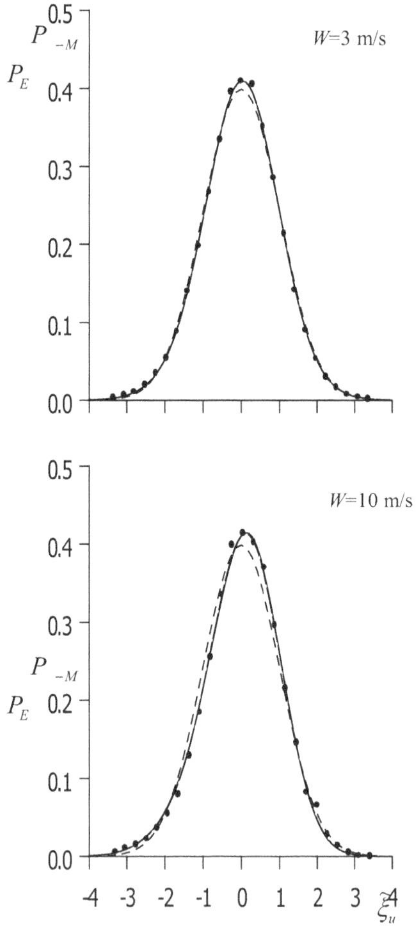

Fig. 2.11 Model P_{N-M} (solid line) and empirical P_E (points) probability density functions of upwind component of sea surface slope, Gauss distribution is dashed line

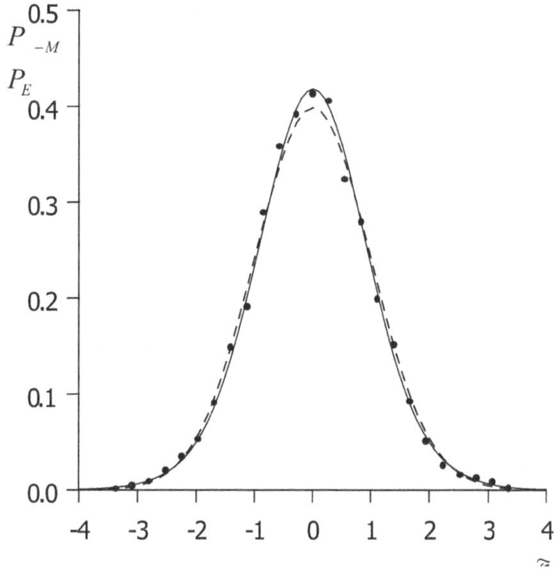

Fig. 2.12 Model P_{N-M} (solid line) and experimental P_E (points) functions of probability density of crosswind component of sea surface slope, Gauss distribution is dashed line

The models presented in Fig. 2.11 are P_{N-M} built with the following parameters: $\varepsilon = 0.61$; при $W = 3$ м/с $m_1 = 0.0542, m_2 = -0.0847, \sigma_1 = 0.8807, \sigma_2 = 1.1572$; при $W = 10$ м/с $m_1 = 0.2857, m_2 = -0.4469, \sigma_1 = 0.8239, \sigma_2 = 1.0839$.

The skewness and kurtosis of the crosswind component of the slope do not depend on wind speed. From Fig. 2.12 we can see that the probability density function obtained in the experiments does not correspond to the Gaussian distribution; in the peak area the values P_E are noticeably higher. At the same time, the model P_{N-M} is in good agreement with model P_E . The parameters of the model shown in Fig. 2.12 have the following values P_{N-M} : $\varepsilon = 0.61, \sigma_1 = 0.8414, \sigma_2 = 1.269$.

Comparison with the data of the field measurements shows that the model (2.19) describes well the real distributions of the upwind and crosswind components of the slope at $\tilde{\xi}_u < 3.5$, $\tilde{\xi}_c < 3.5$. It has no disadvantages typical of models based on truncated Gram-Charlier series. Its own disadvantage is the big volume of calculations.

The question of the probability of large slope needs further analysis. It is necessary to develop special methods aimed at investigation of the "tails" of the slope components distribution and to carry out corresponding field experiments.

2.9 Sea Surface Slope Module

When analyzing quasi-vertical laser data, it is often assumed that its slope are subject to Gauss distribution [7]. Let us consider the correctness of this assumption. We will assume that the average values of the slope component are equal to zero and the correlation coefficient between ξ_u and ξ_c equal to zero. In this case two-dimensional probability density ξ_u and ξ_c is described by the expression:

$$P(\xi_u, \xi_c) = \frac{1}{2\pi\sqrt{\overline{\xi_u^2}\,\overline{\xi_c^2}}} \cdot \exp\left[-\frac{1}{2} \cdot \left(\frac{\xi_u^2}{\overline{\xi_u^2}} + \frac{\xi_c^2}{\overline{\xi_c^2}}\right)\right] \tag{2.32}$$

Let's move from Cartesian to polar coordinate system. Taking into account (2.2), and considering that two-dimensional probability densities of slope in the Cartesian and polar systems of coordinates $P(\xi_u, \xi_c)$ and $P(\xi_m, \alpha)$ binding by the equation

$$P(\xi_m, \alpha) = \frac{\partial(\xi_u, \xi_c)}{\partial(\xi_m, \alpha)} P(\xi_u, \xi_c) \tag{2.33}$$

where $\frac{\partial(\xi_u, \xi_c)}{\partial(\xi_m, \alpha)} = \xi_m$—Jacobian is, we get it

$$P(\xi_m, \alpha) = \frac{\xi_m}{2\pi\sqrt{\overline{\xi_u^2}\,\overline{\xi_c^2}}} \cdot \exp\left[-\frac{1}{2} \cdot \left(\frac{\xi_m^2 \cdot \cos^2\alpha}{\overline{\xi_u^2}} + \frac{\xi_m^2 \cdot \sin^2\alpha}{\overline{\xi_c^2}}\right)\right] \tag{2.34}$$

The expression (2.34) after integration by angle α allows determining the probability density function ξ_m. If you introduce normalization $\tilde{\xi}_m = \frac{\xi_m}{\sqrt{\overline{\xi_u^2} + \overline{\xi_c^2}}}$, it is not difficult to show that in this case the only parameter that defines the distribution type $\tilde{\xi}_m$ is the three-dimensional indicator γ

$$P\left(\tilde{\xi}_m\right) = \tilde{\xi}_m\left(\gamma + \gamma^{-1}\right)\exp\left[-\frac{\tilde{\xi}_m^2\left(\gamma + \gamma^{-1}\right)}{4}\right] I_0\left[\frac{\tilde{\xi}_m^2\left(\gamma^{-2} - \gamma^2\right)}{4}\right] \tag{2.35}$$

where I_0—the Bessel function is zero order.

Let's analyze the slope module distribution ξ_m in two limit cases, when $\gamma = 1$ and when $\gamma = 0$. If $\gamma = 1$, then the slope module is subject to the Relay distribution. Assume that $\sigma_0 = \sqrt{\overline{\xi_u^2}} = \sqrt{\overline{\xi_c^2}}$ then the slope module distribution can be represented as follows

$$P_R(\xi_m) = \frac{\xi_m}{\sigma_0^2} \cdot \exp\left[-\frac{1}{2} \cdot \frac{\xi_m^2}{\sigma_0^2}\right] \tag{2.36}$$

The variance σ_0^2 and variance of the slope module $\overline{\xi_m^2}$ are related by the

$$\overline{\xi_m^2} = \left(2 - \frac{\pi}{2}\right)\sigma_0^2 \tag{2.37}$$

The average value of the slope module $\bar{\xi}_m$ is determined by the expression

$$\bar{\xi}_m = \sqrt{\frac{\pi}{2}} \cdot \sigma_0 \tag{2.38}$$

where the coefficient of variation comes from

$$\sqrt{\overline{\xi_m^2}}/\bar{\xi}_m = 0.52 \tag{2.39}$$

The skewness and kurtosis of the Relay distribution are equal $A_R = 0.631$ and $E_R = 0.245$ accordingly.

The situation that $\gamma = 0$ corresponds to a two-dimensional wave field. In this case, the slope module is subject to the distribution of the random value module, distributed according to the normal law with the mathematical expectation equal to zero

$$p_{mN}(\xi_m) = \frac{2}{\sqrt{2\pi\,\overline{\xi_u^2}}} \cdot \exp\left(-\frac{\xi_m^2}{2\,\overline{\xi_u^2}}\right) \tag{2.40}$$

For the variance and mean value of the slope module, the ratios are fair

$$\overline{\xi_m^2} = \frac{\pi - 2}{\pi}\overline{\xi_u^2} \tag{2.41}$$

$$\bar{\xi}_m = \sqrt{\frac{2}{\pi}\overline{\xi_u^2}} \tag{2.42}$$

of which it follows,

$$\sqrt{\overline{\xi_m^2}}/\bar{\xi}_m = 0.76 \tag{2.43}$$

Skewness and kurtosis of the slope module have the following values $A_{mN} = 0.995$ and $E_{mN} = 0.869$.

Calculated from the measurement data the values of the three-dimensional indicator γ lie in the range from 0.66 to 0.95 [34]. If the slope components ξ_u and ξ_c are subordinate to the distribution (2.32), the values of the skewness and kurtosis of the slope module should have values lying in the interval between the distribution coefficients (2.36) and (2.40), closer to the estimates of coefficients corresponding to the distribution (2.36).

Experimental coefficients of skewness and kurtosis of the sea surface slope module determined at different wind speeds are presented in Fig. 2.13. Practically all estimates of the coefficient A_m lie in the interval between A_R and A_{mN}. However, their distribution within this interval does not correspond to what could be expected from the three-dimensional indicator data analysis, assuming that the slope distribution is

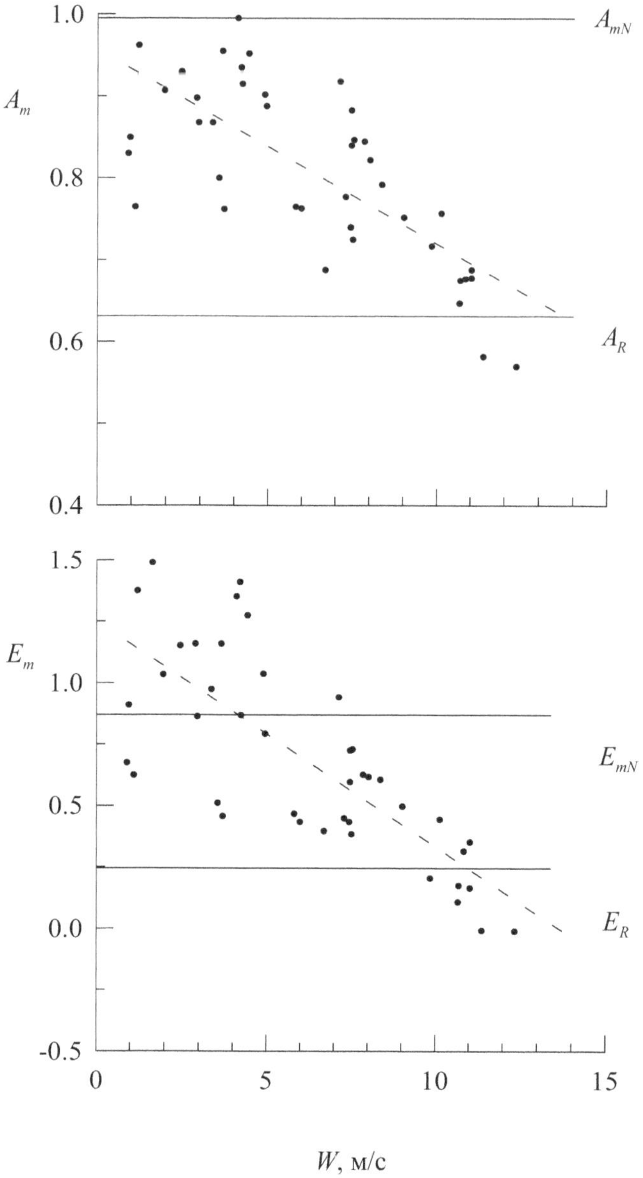

Fig. 2.13 Dependence of skewness A_m and kurtosis E_m of the sea surface slope module on wind speed W. The dashed lines show the regression (2.44) and (2.45) respectively

described by the Gauss distribution. The estimates A_m are not grouped in close proximity A_R, there is a clear dependence A_m from wind speed, which was not observed in the three-dimensional indicator γ (see Fig. 2.6).

The skewness dependence of the slope module on wind speed is approximated by the linear regression equation

$$A_m = -0.025 \cdot W + 0.97 \quad (2.44)$$

with standard deviation ± 0.068. The correlation coefficient of parameters A_m and W is—0.79.

Kurtosis E_m vary widely. At wind speed less than 4 m/s and at wind speed greater than 11 m/s the values go E_m beyond the area which is limited by parameters E_R and E_{mN}. For the kurtosis, as well as for skewness, there is a well-defined dependence on wind speed. This dependence is described by the linear equation

$$E_m = -0.091 \cdot W + 1.23 \quad (2.45)$$

with standard deviation ± 0.23. Correlation coefficient of parameters E_m and W is -0.82.

Another parameter that characterizes the slope module distribution is the coefficient of variation $\sqrt{\overline{\xi_m^2}}/\bar{\xi}_m$ (Fig. 2.14). With increasing wind speed its value on average decreases. The correlation coefficient between the parameters $\sqrt{\overline{\xi_m^2}}/\bar{\xi}_m$ and W is -0.54. The regression equation, describing the relationship between the parameters $\sqrt{\overline{\xi_m^2}}/\bar{\xi}_m$ and W, has the form:

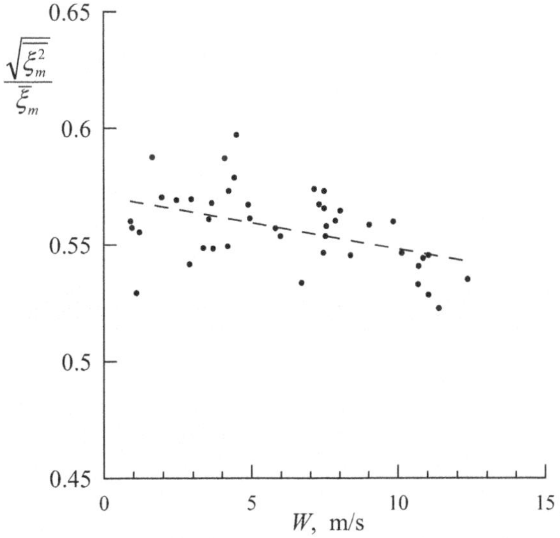

Fig. 2.14 Dependence of variation coefficient $\sqrt{\overline{\xi_m^2}}/\bar{\xi}_m$ on wind speed W. The bar line shows the regression (2.46)

$$\sqrt{\overline{\xi_m^2}}/\bar{\xi}_m = -0.0023 \cdot W + 0.57 \tag{2.46}$$

Standard deviation is ±0.012.

The coefficient of variation values determined in the experiments lie in the range of 0.52–0.59. The lower limit of this range is the same as the value corresponding to the Relay distribution, the upper limit is much lower than the value 0.76 corresponding to the distribution (2.40).

Thus, all parameters characterizing the statistical distribution of the sea surface slope module (skewness, kurtosis, coefficient of variation) turned out to be dependent on wind speed. If the two-dimensional probability density of the slope components ξ_u and ξ_c is described by the expression (2.32), then, as shown above, the only parameter on which the skewness and kurtosis of the slope module depend is the three-dimensional indicator γ. Let's consider how the parameters A_m, E_m and γ are related, in the real wave field (Fig. 2.15).

When constructing Fig. 2.15, the estimates of the three-dimensional indicator obtained at wind speed less than 4 m/s were highlighted. As will be shown below, the weak wind regime has a number of features associated with the appearance of slicks on the sea surface. From Fig. 2.15 it follows that the variations of the three-dimensional indicator observed in field conditions do not have a noticeable effect on the skewness and kurtosis of the slope module.

The nature of changes in the mean slope modulus $\bar{\xi}_m$ and standard deviation $\sqrt{\overline{\xi_m^2}}$ with increasing wind speed are shown in Fig. 2.16. The dependencies on wind speed,

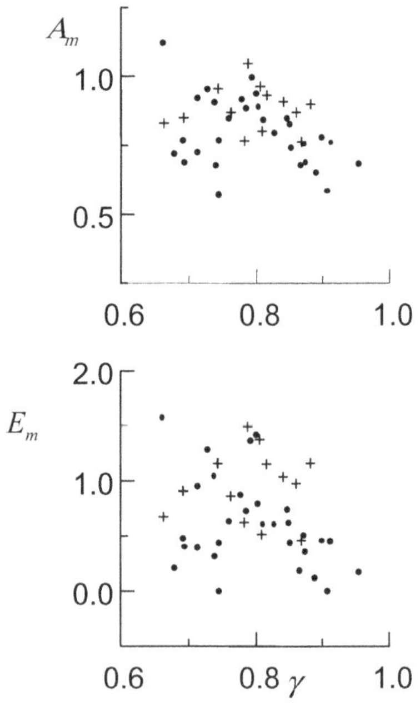

Fig. 2.15 Dependence of skewness A_m and kurtosis of the slope module E_m on the three-dimensional indicator γ. Symbol + indicates estimates obtained at wind speeds below 4 m/s

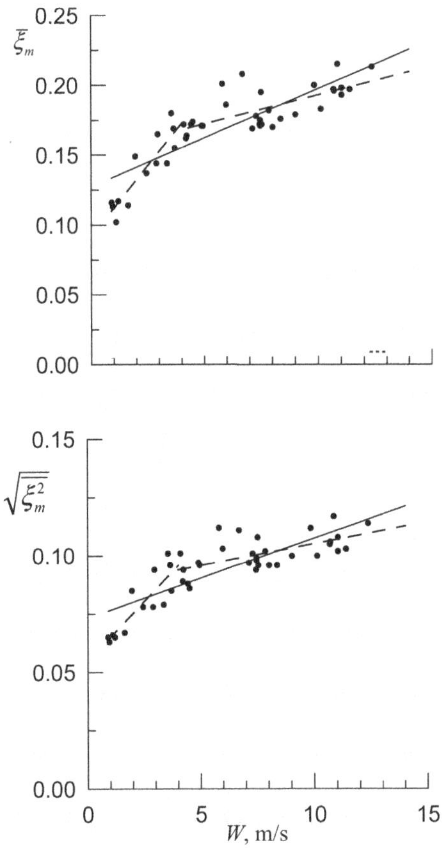

Fig. 2.16 Dependencies of parameters $\bar{\xi}_m$ and σ_m wind speed W. Solid lines—regressions (2.47) and (2.48), accordingly, dashed lines—regressions (2.49) and (2.50) accordingly

for both parameters are similar, with increasing wind speed their values are as follows

$$\bar{\xi}_m = 6.9 \times 10^{-3} \cdot W + 0.129 \pm 0.015 \quad (2.47)$$

$$\sqrt{\overline{\xi_m^2}} = 3.41 \times 10^{-3} \cdot W + 0.074 \pm 8.8 \cdot 10^{-3} \quad (2.48)$$

Regression Eqs. (2.47) and (2.48) are constructed for the whole range of wind speeds in which measurements were made. Correlation coefficients $\bar{\xi}_m$ and $\sqrt{\overline{\xi_m^2}}$ for parameters and with wind speed are 0.85 and 0.80, respectively. It also follows from Fig. 2.16 that the slope characteristics change differently in different ranges

$$\bar{\xi}_m = \begin{cases} 20.2 \times 10^{-3} \cdot W + 0.0917, & \text{W} \leq 4 \text{ m/s} \\ 5.03 \times 10^{-3} \cdot W + 0.143, & \text{W} > 4 \text{ m/s} \end{cases} \quad (2.49)$$

$$\sqrt{\overline{\xi_m^2}} = \begin{cases} 11.3 \times 10^{-3} \cdot W + 0.0513, & \mathrm{W} \leq 4 \text{ m/s} \\ 1.89 \times 10^{-3} \cdot W + 0.0863, & \mathrm{W} > 4 \text{ m/s} \end{cases} \tag{2.50}$$

The fastest growth $\bar{\xi}_m$ and $\sqrt{\overline{\xi_m^2}}$ is observed at wind speeds of up to 4 m/s. In the linear regression equation at $\bar{\xi}_m$, the coefficients at W calculated for two wind velocity ranges, $W \leq 4$ m/s and $W > 4$ m/s, differ fivefold. Even greater difference is observed for the parameter $\sqrt{\overline{\xi_m^2}}$, its growth rate differs eight times.

Summarizing the analysis of data from in situ studies of the sea surface slope module variability conducted on the oceanographic platform, we note the following. Under weather conditions varying from calm to wind at a speed of 14 m/s, the coefficients characterizing the slope module distribution vary widely: from A_m 0.57 to 0.99; E_m—from −0.01 to 1.38; $\sqrt{\overline{\xi_m^2}}/\bar{\xi}_m$—from 0.52 to 0.59. The values of the coefficients A_m, E_m and $\sqrt{\overline{\xi_m^2}}/\bar{\xi}_m$ have a high level of correlation with wind speed (correlation coefficients, respectively, are equal to: −0.79, −0.82, −0.54). At any wind speeds, there was no situation when all three coefficients A_m, E_m and $\sqrt{\overline{\xi_m^2}}/\bar{\xi}_m$ at the same time approached the values corresponding to the Relay distribution (2.36) or the random value module distribution, distributed by the normal law (2.40).

Variability of sea surface characteristics in low winds

Weak winds are most favorable for monitoring the processes occurring in the upper ocean and the atmosphere by remote sensing. Studies of sea surface roughness variations have shown that not only smoothing bands but also reinforced roughness bands often appear on the surface as a result of internal waves. Studies carried out from a moving vessel using contact equipment in the coastal part of the Black Sea have shown that in these situations, the short-wave energy in the slick decreases by 3–6 dB, immediately after the slick increases by 2–4 dB compared to the background level and then decreases again to the background level [35]. Similar phenomena are observed in the open ocean [25].

2.10 Statistics of Sea Surface Slope in Areas of Slicks and Ripples

On the sea surface, there are often areas where short waves are partially or completely suppressed, these areas are commonly called slices. The physical mechanisms causing the appearance of slicks are very diverse [27]. Slicks may occur as a result of natural processes or anthropogenic impacts on the marine environment. The appearance of natural slicks is caused by surface current variations created by internal waves, wind speed reduction below the threshold value at which ripples are generated, Langmuir circulation, films of biogenic substances, etc. Anthropogenic slicks are usually produced by an oil spill. Slicks created by different physical mechanisms have different sizes, configuration and speed of movement that, usually, allows to define the nature of their origin on the basis of visual observations.

Data from sea surface slope measurements performed in situations where natural slicks and ripple areas were simultaneously visually observed on the surface were divided into two groups [35]. The first group included statistical moments of the two-dimensional distribution of slope obtained in situations where the slopemeter was in the slick area, these data are presented in Table 2.3. The second group includes statistical moments obtained in ripple zones between slicks (Table 2.4). The fifth column of Table 2.3, which contains data on statistical moments of slope in artificial slick, contains data from the work [2]. The artificial slick was formed by a spill on the sea surface mixture consisting of 40 percent used crankcase oil, 40% Diesel oil, and 20% fish oil.

The data presented in the tables allow to reveal characteristic features that distinguish the statistics of slope in the areas of slicks and ripples. Coefficients of proportionality in regression equations describing the dependence of variance of upwind and crosswind components of slope on wind speed, for smoothed and rough surfaces were close to each other. These coefficients are approximately 3–4 times higher than in similar regression equations for pure wind waves in moderate and strong winds.

Data from sea surface slope measurements suggest that in weak winds, in situations where slicks are observed on the sea surface, the structure of the sea surface changes much more drastically than in the subsequent stages of wind wave development [12]. The conclusion is also confirmed by the fact that in addition to variance, some statistical parameters undergo characteristic changes in the transition from one hydrodynamic situation to another. The effect is clearly shown in slicks (statistical moments of slope are given in Table 2.3): at calm the coefficients of kurtosis $\lambda_{0,4}$ and

Table 2.3 Statistical characteristics of sea surface slope in the slick zone

Parameter	Smooth slices, $W <$ 0.8 m/s	Slices from internal waves, $0.8 < W <$ 1.7 m/s	Slices from internal waves, $1.7 < W$ < 3.0 m/s	Artificial slices [21], $1.8 < W <$ 10.8 m/s
1	2	3	4	5
$\overline{\xi_c^2} \times 10^3$	1.72 ± 0.5	1.05 ± 0.2	$-10.51 + 6.92\,W$; ± 0.94; r $= 0.96$	$3.3 + 0.8\,W$; ± 2; r $= 0.80$
$\overline{\xi_u^2} \times 10^3$	2.6 ± 1.2	1.9 ± 0.4	$-12.63 + 8.73\,W$; ± 1.5; r $= 0.94$	$4.6 + 0.75\,W$; ± 2; r $= 0.75$
$\lambda_{0,3}$	-0.13 ± 0.11	-0.19 ± 0.22	-0.09 ± 0.24	0.02 ± 0.05
$\lambda_{1,2}$	0.16 ± 0.05	0.06 ± 0.01	0.05 ± 0.14	0
$\lambda_{2,1}$	0.14 ± 0.03	0.01 ± 0.02	0.03 ± 0.07	0.00 ± 0.02
$\lambda_{3,0}$	-0.15 ± 0.20	-0.06 ± 0.20	0.01 ± 0.21	0
$\lambda_{0,4}$	0.14 ± 0.39	0.15 ± 0.60	0.32 ± 0.51	0.26 ± 0.31
$\lambda_{2,2}$	0.04 ± 0.19	-0.06 ± 0.13	0.08 ± 0.16	0.10 ± 0.05
$\lambda_{4,0}$	0.14 ± 0.52	-0.02 ± 0.30	0.32 ± 0.45	0.36 ± 0.24

Table 2.4 Statistical characteristics of sea surface slope in the ripple zone

Parameter	Jubilee between slices $0.8 < W < 2.0$ m/s	Jubilee between slices $2.0 < W < 3.3$ m/s	A pure windstorm $4.0 < W < 11.0$ m/s
$\overline{\xi_c^2} \times 10^3$	$-4.27 + 6.09\,W;\ \pm 1.71;\ \mathrm{r} = 0.88$	8.56 ± 4.35	$-.46 + 2.61\,W;$ $\pm 3.1;\ \mathrm{r} = 0.82$
$\overline{\xi_u^2} \times 10^3$	$-7.00 + 9.96\,W;\ \pm 2.35;\ \mathrm{r} = 0.84$	13.85 ± 5.45	$2.82 + 3.02\,W;$ $\pm 3.01;\ \mathrm{r} = 0.86$
$\lambda_{0,3}$	0.06 ± 0.23	-0.04 ± 0.21	-0.02 ± 0.27
$\lambda_{1,2}$	0.00 ± 0.12	-0.03 ± 0.12	-0.01 ± 0.12
$\lambda_{2,1}$	0.02 ± 0.10	0.02 ± 0.09	-0.00 ± 0.09
$\lambda_{3,0}$	0.07 ± 0.13	0.11 ± 0.19	-0.10 ± 0.19
$\lambda_{0,4}$	0.40 ± 0.42	0.37 ± 0.41	0.48 ± 0.51
$\lambda_{2,2}$	0.23 ± 0.32	0.09 ± 0.15	0.09 ± 0.13
$\lambda_{4,0}$	0.33 ± 0.48	0.31 ± 0.41	0.35 ± 0.42

$\lambda_{4,0}$ are equal to 0.14, and in light wind they increase to 0.32. This is only slightly less than in the same wind speeds in ripple zones between slicks, where $\lambda_{0,4} = 0.37$ and $\lambda_{4,0} = 0.31$. The observed changes in sea surface characteristics can be explained by the fact that as the wind increases the nonlinearity of waves increases [32], and as a result the deviation from the Gaussian distribution increases.

The symbol in r Tables 2.3 and 2.4 indicates the correlation coefficient.

The data in Table 2.3 show that the process of changing the structure of the sea surface does not develop monotonically. The sharpest changes in most of the slope distribution parameters occur in the transition from a smooth surface to a rough one, and further changes are mainly due to the relatively slow growth of slope variance with increasing wind speed. The pure wind stage starts with wind speeds of 4–5 m/s, the distribution parameters for this stage (in the range of $4 \le W \le 11$ m/s) are given in the last column of Table 2.4.

2.11 Characteristics of Sea Surface Slope in Natural and Artificial Slicks

Along with natural slices, it is often necessary to observe technogenic slices on the sea surface. As a rule, anthropogenic slicks are caused by sea pollution with oil products. The effect of ripple suppression by surfactants is widely used to control pollution by remote sensing devices installed on spacecraft [9, 26].

We will compare our registered surface structure changes in natural slicks created by internal waves or calm weather conditions with published data on statistical slope moments in oil slicks. For this purpose, we use the published data on the slope

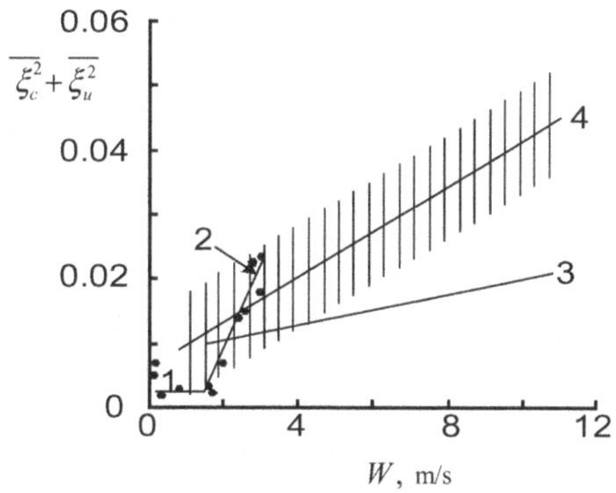

Fig. 2.17 Behaviour of total variance of slope $\left(\overline{\xi_c^2}+\overline{\xi_u^2}\right)$ in slices. Points are measurement data in natural slicks; straight lines 1, 2 are their linear approximations built at different wind speeds; line 3 is linear regression for oil slicks [2]; line 4 is linear regression of slope of rough surface [35] (shaded area of standard deviations)

characteristics in artificial slicks, which were formed by spill on the sea surface of oil products in a mixture with fish oil [2].

The differences between the statistical characteristics of slope in natural and oil slicks (they are presented in Table 2.3) are significant [12]. At wind velocities above 1.7 m/s, the growth of slope variance in natural slicks is approximately 10 times faster than in artificial slicks (Fig. 2.17). Oil slices may continue to exist even at wind speeds of 10–11 m/s, while natural slices associated with internal waves tend to disappear at $W = 4$–5 m/s.

The values of the third and fourth order moments of the two-dimensional distribution of slope in artificial and natural slicks turned out to be almost the same and independent of wind speed. It means that statistics of artificial and natural slicks differ from each other only by the growth rate of slope variance with wind gain. Comparison of the data in Tables 2.3 and 2.4 shows that the rate of change in the slope variance with wind gain in artificial slicks was much lower than in the rough surface areas.

The reasons for the difference in changes with increasing wind speed of the variance of slope in natural and oil slicks are probably due to differences in the physics of their formation. Natural slicks were of dynamic origin (calm slicks that appear as the wind friction tension decreases, or slicks caused by orbital currents in internal waves), while artificial "film" slicks appear as a result of changes in surface tension [16].

2.12 Conclusion

The slope statistics of the sea surface, determined from the measurement data at the point by a laser slopemeter and from photographs of sun's glitter were close. This indicates that it is possible to use the measurement data from the laser slopemeter of the local wave field characteristics to build models describing the formation of sea surface images during remote sensing. The proximity of the slope characteristics determined from optical images of the sea surface and from point measurements also indicates the validity of the ergodic hypothesis for a process such as the sea wind wave.

The analysis of possibilities and limitations of description of sea surface slope distribution by means of truncated Gram-Charles series has been carried out. In the area of large values of slope, which are of interest for remote sensing tasks, negative values of the models of the probability density function appear. The boundary of negative values appearance has been calculated using the data of direct measurements of slope.

It is shown that the real distribution of the slope module differs significantly from the model one obtained under the assumption that the slope components are subject to the Gaussian distribution. Regression dependencies describing the relationship of slope module cumulants to the wind are constructed by the laser slopemeter.

Variability of slope characteristics in situations where slicks and ripples are present on the sea surface simultaneously has been studied. It is shown that the slope variance in natural slicks in weak winds can have values of the same order as those of wind waves, but in slicks the probability distribution is close to the Gauss distribution, and in wind waves with wind reinforcement the deviations from it increase. The process of sea surface structure change is non-monotonous. The most abrupt changes of slope aggregates are observed in the transition from a smooth surface to a rough one, further changes are mainly reduced to relatively slow growth of slope variance with increasing wind speed.

It is established that in oil and natural slicks the values of third and fourth order cumulants of two-dimensional distribution of slope are close to each other and do not depend on wind speed. Statistics of oil and natural slicks differ from each other only by the dynamics of changes in the slope variance. At wind speed above 1.7 m/s, the growth of the slope variance in natural slicks with the growth of wind speed increases approximately 10 times faster than in artificial slicks.

The increased attention paid to the study of sea surface slope is due to the fact that they are the main characteristic of the sea surface, when modeling the interaction processes of electromagnetic waves with the sea surface. The problem of studying the slope variability is that the main contribution to their variance is made by short waves, whose characteristics are technically difficult to measure under field conditions. The cycle of experimental studies conducted on the oceanographic platform described in Chap. 2 provides a notable contribution to solving this problem.

References

1. Chu X, He Y, Chen G (2012) Asymmetry and anisotropy of microwave backscatter at low incidence angles. IEEE Trans. Geosci. Remote Sens. 50:4014–4024
2. Cox C, Munk W (1954) Measurements of the roughness of the sea surface from photographs of the sun glitter. J. Opt. Soc. Am. 44(11):838–850
3. Danilytchev MV, Kutuza BG, Nikolaev AG (2009) The application of sea wave slope distribution empirical dependences in estimation of interaction between microwave radiation and

rough sea surface. IEEE Trans. Geosci. Remote Sens. 47(2):652–661. https://doi.org/10.1109/tgrs.2008.2004410

4. Ebuchi N, Kizu S (2002) Probability distribution of surface wave slope derived using Sun glitter images from geostationary meteorological satellite and surface vector winds from scatterometers. J. Oceanogr. 58:477–486
5. F.G. Bass, I.M. Fuks, *Wave scattering by statistically rough surface* (Pergamon, 1979), 540 p
6. F.M. Bréon, N. Henriot, Spaceborne observations of ocean glint reflectance and modeling of wave slope distributions. J. Geoph. Res. Oceans. **111**(C06005) (2006). https://doi.org/10.1029/2005jc003343
7. Hu Y, Stamnes K, Vaughan M, Pelon J, Weimer C, Wu D, Cisewski M, Sun W, Yang P, Lin B, Omar A, Flittner D, Hostetler C, Trepte C, Winker D, Gibson G, Santa-Maria M (2008) Sea surface wind speed estimation from space-based lidar measurements. Atmos. Chem. Phys. 8(13):3593–3601
8. Hughes BA, Grant HL, Chappell RWA (1977) A fast response surface-wave slope meter and measured wind-wave components. Deep-Sea Res. 24(12):1211–1223
9. Ivanov A.Yu. Slicks and film formations in SAR data// Issledovanie Zemli iz kosmosa, 2007, No. 3, pp. 73–96
10. Karaev VYu, Kanevsky MB, Meshkov EM, Titov VI, Balandina GN (2008) Measurement of the variance of water surface slope by a radar: verification of algorithms. Radiophys. Quantum Electron. 51(5):360–371. https://doi.org/10.1007/s11141-008-9042-6
11. Kendall MJ, Stewart A (1958) *The Advanced Theory of Statistics. Vol. I. Distribution theory* (Butler & Tanner Ltd., London), 675 p
12. Khristoforov GN, Zapevalov AS, Babii MV (1992) Measurements of sea-surface roughness parameters during the transition from calm to wind generated wave // Izvestiya Akademii Nauk Fizika Atmosfery i Okeana 28(4):424–431
13. Khristophorov GN, Zapevalov AS, Babiy MV (1992a) Statistics of sea-surface slope for different wind speeds. Okeanologiya 32(3):452–459
14. Khristoforov GN, Zapevalov AS, Babii MV (1992b) Measurements of sea-surface roughness parameters during the transition from calm to wind generated wave. Izvestiya Akademii Nauk Fizika Atmosfery i Okeana 28(4):424–431
15. Kistovich AV, Pokazeev KV, Chaplina TO (2020) *Physics of the Sea* (Moscow, Yureit Publishing House), 336 p
16. Kistovich AV, Pokazeev KV (2010) *Actual problems of ocean surface physics* (Moscow, Physics Faculty of Moscow State University), 232 p
17. Lebedev NE, Aleskerova AA, Plotnikov EV (2016) The development of optical methods for sea surface slope measurement. Sovremennye problemy distantsionnogo zondirovaniya Zemli iz kosmosa 13(3):136–149
18. Lebedev NE, Pustovoitenko VV, Pokazeev KV, Melnikova ON (2014) Simulation of the sea surface bidirectional reflectance distribution function // Sovremennye Problemy Distantsionnogo Zondirovaniya Zemli iz Kosmosa 11(3):310–320
19. Longuet-Higgins MS (1957) The statistical analysis of a random, moving surface. Philos. Trans. R. Soc. A Math. Phys. Eng. Sci. 249(966):321–387. https://doi.org/10.1098/rsta.1957.0002
20. Longuet-Higgins MS (1982) On the skewness of sea-surface slope. J. Phys. Oceanogr. **12**, 1283–1291https://doi.org/10.1175/1520-0485(1982)012
21. P. Chen, Q. Yin, P. Huang, Effect of non-Gaussian properties of the sea surface on the low-incidence radar backscatter and its inversion in terms of wave spectra by an ocean wave. Chin. J. Oceanol. Limnol. **33**(5), 1142–1156 (2015). https://doi.org/10.1007/s00343-015-4191-x
22. Plant WJ (2002) A stochastic, multiscale model of microwave backscatter from the ocean // J. Geophys Res 107(C9):3120 https://doi.org/10.1029/2001JC000909
23. Pokazeev KV (1988) Modulation of wind ruffles with long waves, *Proceedings of GOIN* 188:121–125
24. Pustovoytenko VV, Lebedev NE (2015) Comparison of sea surface slope statistical moments obtained by means of optical scanners and laser inclinometers. Sovremennye Problemy Distantsionnogo Zondirovaniya Zemli iz Kosmosa 12(1):102–109

25. Serebryany AN (2012) Slick- and suloy generating processes in the sea. Internal waves. Sovremennye problemy distantsionnogo zondirovaniya Zemli iz kosmosa. 9(2): 275–286
26. Shcherbak SS, Lavrova OYu, Mityagina MI, Bocharova TYu, Krovotyntsev VA, Ostrovskii AG (2008) Multisensor satellite monitoring of seawater state and oil pollution in the northeastern coastal zone of the Black Sea. Int. J.Remote Sens. 29(21):6331–6345
27. Shmirko KA, Pavlov AN, Dubinkina ES, Konstantinov OG (2017) Peculiarities of slick formation on the sea surface. Izvestiya. Atmos. Oceanic Phys. 53(1):95–101
28. Su W, Charlock TP, Rutledge K (2002) Observations of reflectance distribution around sunglint from a coastal ocean platform. Applied Optics 41(35):7369. https://doi.org/10.1364/ao.41.007369
29. Tatarskii VI (2003) Multi-Gaussian representation of the Cox–Munk distribution for slopes of wind-driven waves // J. of Atmos Oceanic Technol 20:1697–1705
30. Zapevalov AS (2009) Bragg scattering of centimeter electromagnetic radiation from the sea surface: The effect of waves longer than Bragg components. Izvestiya Atmos. Ocean Phys. 45(2):253–261
31. Zapevalov AS (2018) Determination of the statistical moments of sea-surface slope by optical scanners. Atmos. Oceanic Opt. 31(1):91–95. https://doi.org/10.1134/S1024856018010141
32. Zapevalov AS, Bol'shakov AN, Smolov VE (2011) Simulating of the probability density of sea surface elevations using the Gram-Charlier series. Oceanology 51(3):406–413
33. Zapevalov AS, Ratner YuB (2003) Analytic model of the probability density of slope of the sea surface. Phys. Oceanogr. 13(1):1–13
34. Zapevalov AS (2002) Statistical characteristics of the moduli of slope of the sea surface. Phys. Oceanogr. 12(1):24–31
35. Zapevalov AS (2008) Statistical models of the sea surface in problems of acoustic and electromagnetic radiation scattering, Manuscript to claim the academic degree of doctor of physico-mathematical sciences on the speciality 04.00.22 – geophysics (Marine Hydrophysical Institute of the National Academy of Sciences of Ukraine, Sevastopol)
36. Zhang H, Wang M (2010) Evaluation of sun glint models using MODIS measurements. J. Quant. Spectro. Rad. Trans. 111(3):492–506

Chapter 3
On the Dispersion Relation of Sea Waves

3.1 Introduction

The multi-channel optical systems installed on modern satellites allow receiving images of the sea surface with high spatial resolution, which allows restoring its relief on scales smaller than the dominant waves [19]. Small temporal lag image sequences make it possible to determine the phase velocity of surface waves and, based on these data, to estimate the current velocity and bathymetry of the coastal zone [20, 47]. Navigational and specialized ship and shore-based radars are used to provide operational information on the current field [9, 13]. Signals from the ridges of long surface waves are clearly visible on radar images, which makes it possible to use series of sequential images to determine current velocity [14].

When interpreting remote sensing data, the question of spatial-temporal relationships in the wave field, nonlinear effects and dispersion relation disturbances becomes very important. The remote methods of determining the velocity of the current are based on the assumption that in the absence of a current for gravity waves in deep water the dispersion relation is fair [7]

$$\omega^2 = gk \tag{3.1}$$

where ω is an angular frequency; g is gravitational acceleration; k is wave number.

For the first time, the dispersion relation disorder in field conditions for gravity waves in deep water was experimentally shown [44]. This result was obtained on the basis of measurements of an array of spaced wave sensors mounted on fixed supports. The dispersion relation disturbances were later confirmed in a series of laboratory and field experiments [26, 27, 38, 39, 50]. A number of physical mechanisms lead

A. Zapevalov et al., *Simulation of the Sea Surface for Remote Sensing*, Springer Oceanography,
https://doi.org/10.1007/978-3-030-58752-9_3

to dispersion relation disorders. These include: kinematic nonlinearity, presence of bound components, wave-wave interactions, impact of longer waves on short waves, etc. [11, 18, 22, 34, 35].

One of the important characteristics of the wave field, describing its space-time structure, is the spreading function that describes angular distribution of wave energy. The bulk of information on the frequency-angle characteristics of surface waves was obtained with the help of wave buoys of "heavy-pitch-roll" type [4, 32, 40, 42, 45]. Significantly fewer studies have been carried out using a high resolution directional array consisting of string sensors [2, 10, 52]. Multi-point measurements provide more detailed estimates of frequency and angle characteristics, but are technically more complex and require significantly higher financial costs.

This chapter analyses the spatio-temporal structure of the sea surface wave from in situ measurements and builds models to describe it.

3.2 Wave Measurement Equipment and Measurement Conditions

The results of field studies presented in this chapter are obtained from the oceanographic platform of the Marine Hydrophysical Institute. The oceanographic platform provides unique opportunities for complex studies of processes occurring near the water-air interface and at the interface itself [21, 30]. It is installed on the Black Sea near the Southern coast of Crimea, near the village of Katsiveli. The minimum distance from the platform to the shore is about 500 m. The depth at the place where the platform is installed is 30 m. For characteristic wavelengths in the measurement area, this depth corresponds to the "deep water" condition, and the influence of the seabed on wind waves and swell can be neglected.

The requirements to the waveguide equipment during contact measurements in support of remote sensing tasks are primarily determined by the scale of surface waves scattering electromagnetic radiation. For satellite oceanography tasks, both short-wave wavelength studies corresponding to centimetre and decimetre wavelengths [3] and long-wavelength studies are of interest [5, 51].

String resistive wave gauges are widely used in field research. The devices of this type allow to measure waves in a range of scales from basic energy-bearing waves with lengths about 100 m, up to ripple waves with length about 1 cm [50]. The gauge of this type is a section of uninsulated wire that vertically crosses the water-air interface. Sensors made of nichrome wire were used in experiments on an oceanographic platform. The choice of material for the sensors was determined by the requirements of mechanical strength, anticorrosive properties and high electrical resistance.

A oceanographic platform mounted on the seabed, like any other physical object crossing the ocean-atmosphere boundary, causes disturbances in the surface wave. A number of measures have been taken to avoid, as far as possible, the impact of disturbances generated by the supports of the ocean platform on the measurement results. Wave measurements were made only from the sea side of the platform. Для установки аппаратуры использовались выстрелы, позволяющие удалить от опор области, в которых проводятся волнографические измерения. The flow in the area of the oceanographic platform, as a rule, is directed along the shore, so the influence of the wave disturbances created by the platform supports in the measurement area was insignificant with the chosen arrangement of the equipment.

Several types of tooling were used to accommodate the waveform string sensors. Special tooling, mounted at the end of the six-meter boom, made it possible to install an array of sensors consisting of a waveguide and four differential slopemeters with bases from 1 to 5 cm (two slopemeters each in orthogonal directions). On the same equipment was installed a one-dimensional wave gauge array, consisting of five wave gauges. The minimum distance between the gauges was 8 cm, the maximum—240 cm. The design of the grid attachment made it possible to change its orientation in the azimuthal plane. This made it possible to consistently determine the characteristics of the sea wave, both in the direction of wave propagation and in the orthogonal direction [49].

Another type of used wave grids is the "cross" (two-dimensional array), which allows synchronous measurements of spatial and temporal characteristics in two mutually perpendicular directions. The maximum distance between the gauges was 45 cm, the number of gauges was equal to 5. As with one-dimensional array measurements, the azimuthal orientation of the two-dimensional array could be adjusted to the direction of the waves.

In the case of small waveguide arrays, the design of the stringline sensor rigging had both surface and underwater booms. The booms were attached to a vertically oriented bar. The diameter of the rod was 50 mm. String sensors were stretched between the above and below and were sprung from their lower ends by rubber shock absorbers [50].

For the study of spatial and temporal characteristics on the scale of long dominant waves, portable booms with a length of 6 m were used. They allowed to place the wave gauges in such a way that the radius-vectors binding them were oriented along or across the main wave propagation direction. The use of man-portable booms allowed measurements to be made in storm conditions, with wind speeds up to 25 m/s. The maximum distance between sensors was 20 m [15].

3.3 Measurements of the Quadratic Coherence Function on the Scale of the Dominant Waves

When measuring an array of wave gauges, the spatial and temporal structure of the sea surface elevations is determined by the wave field cross-spectrum $X(f, \vec{L})$.

The level of linear statistical coupling in the spectral region at two points of space $Y_1 = (x_1, y_1)$ and $Y_2 = (x_2, y_2)$ separated by a vector $\vec{L}$ is described by a quadratic coherence function.

$$R^2(f, \vec{L}) = \frac{Co^2(f, \vec{L}) + Q^2(f, \vec{L})}{S_1(f)\ S_2(f)} \tag{3.2}$$

where $f = \omega/(2\pi)$ is frequency; $Co(f, \vec{L}) = \mathrm{Re}\big(\mathrm{X}(f, \vec{L})\big)$ is co-spectrum; $Q(f, \vec{L}) = \mathrm{Im}\big(\mathrm{X}(f, \vec{L})\big)$ is quadrature spectrum; $S_1(f)$ and $S_2(f)$ is surface elevation spectra measured in points Y_1 and Y_2. If the wave field is homogeneous, then equality $S_1(f) = S_2(f) = S(f)$ is fair.

It should be noted that in the theoretical analysis of the wave fields it is more convenient to use an angular frequency ω, while in the analysis of experimental data it is more convenient to use the frequency f. In this chapter, depending on the context, we will use either, ω or f.

To analyze the azimuthal characteristics of the sea surface waves, we introduce the concept of longitudinal and transverse coherence (let us denote respectively the quadratic coherence functions as $R^2_{\uparrow}$ and $R^2_{\perp}$), which we define as coherence along and across the main wave propagation direction. We also introduce the notion of coherence anisotropy

$$A_R(f, L) = \frac{R^2_{\perp}(f, L)}{R^2_{\uparrow}(f, L)}. \tag{3.3}$$

where L—distance between the sensors.

Characteristic changes in the longitudinal and transverse quadratic coherence functions are shown in Fig. 3.1. Measurements were made at low wind, when its speed U was equal to 2 m/s, and at strong wind $U = 10$ m/s. Quadratic coherence functions are constructed on the basis of measurements by resistive string wave gauges, spaced at 0.4 m and 2.4 m. Measurements were made on an oceanographic platform [17]. The values of the quadratic coherence function that exceed the level of 0.2 are significant with the probability of 99.7%.

Dashed vertical lines on each graph show: the peak frequency in the elevation spectrum of the sea surface f_0, as well as the frequency $\sqrt{2}\, f_0$ and $2\, f_0$, respectively. The behavior of the functions $R^2_{\uparrow}(f, L)$ and $R^2_{\perp}(f, L)$ defined for the two orthogonal directions in weak and strong winds is similar, with increasing frequency level of coherence decreases.

Let's analyze how the coherence level changes when the vector $\vec{L}$ angle θ changes. To determine the angular characteristics of surface waves, we use the phase spectrum $\phi(f, L, \theta) = \arctan\big(-\mathrm{Im}(\mathrm{X}(f, L, \theta))/\mathrm{Re}(\mathrm{X}(f, L, \theta))\big)$. The azimuthal dependencies of the coherence level $R^2(f = f_0, L, \theta)$ and the phase shift modulus $|\phi(f = f_0, L, \theta)|$, determined at the frequency of the dominant waves, are shown in Fig. 3.2 [50]. The resulting estimates correspond to the three spacing between gauges,

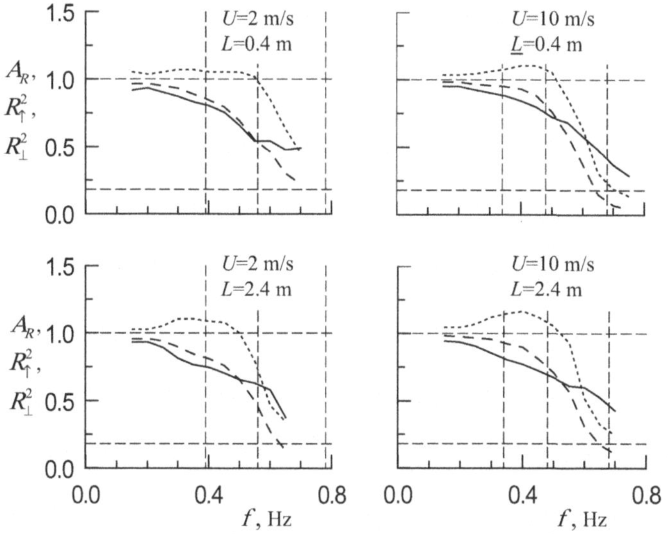

Fig. 3.1 Functions of longitudinal $R^2_{\uparrow}$, transverse $R^2_{\perp}$ square coherence and their anisotropy A_R in strong and weak winds. Symbols —— $R^2_{\uparrow}$; – – $R^2_{\perp}$; ··· A_R

Fig. 3.2 Dependencies of coherence $R^2(f = f_0, \theta)$ and phase shift modulus $|\phi(f = f_0, \theta)|$ on the scale of dominant waves on the azimuth angle θ. Curves 1, 2 and 3 correspond to transducer distances of 35.2, 41.4 and 60 cm

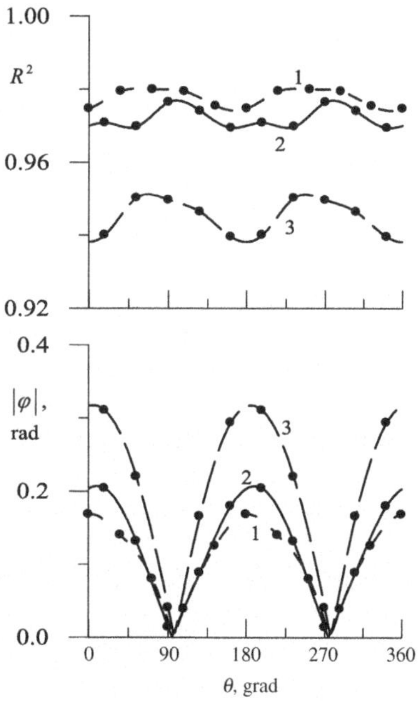

35.2, 41.4 and 60 cm. The maximum phase shift values correspond to the main wave propagation direction, while the minimum values correspond to the orthogonal direction.

In composing Fig. 3.2, the data obtained using an array consisting of six string gauges. Five sensors were located at the tops of a regular pentagon inscribed in a circle with a radius of 35.2 cm, the sixth was in its center. This arrangement of the gauges in the lattice allows to obtain 15 independent spatial and temporal estimates, which are divided into three groups corresponding to three distances between the gauges. For each pair of gauges it is possible to compare two estimations of coherence and phase shift determined for forward and reverse directions, so for each group it is possible to construct 10 estimations evenly distributed in the range 0°—360 in 36° increments.

Despite the fact that the values of the quadratic coherence function change within small limits, it is clearly seen that in the main wave propagation direction the function values $R^2(f = f_0, L, \theta)$ are lower than in the orthogonal direction. The average deviation of the position $\min\left(R^2(f = f_0, L, \theta)\right)$ from the position $\max(|\phi(f = f_0, L, \theta)\,|)$ as well as the position $\max\left(R^2(f = f_0, L, \theta)\right)$ from the position $\max\left(R^2(f = f_0, L, \theta)\right)$ is 11° and 17°, respectively [50].

The experimental manifestations of nonlinearity of wind waves usually include deviations of phase velocity estimates from the values predicted by the linear spectral theory, as well as deviations from the linear dispersion relation of the position of the space-time spectra maxima [11, 46]. Under field conditions, deviations from linear theory are usually observed at values of dimensionless frequency

$$\Omega = f / f_0, \tag{3.4}$$

lying in the range $\Omega \geq 2$.

Based on these results, it could be expected that in the range $1 \leq \Omega < 2$ the quadratic coherence function will correspond to the linear model of the wave field. It could be assumed that the level of coherence between the surface elevations measured at two points of the wave field was mainly determined only by the width of the angular distribution function of the wave energy and the distance between the sensors. In this case, within the limits of linear theory at a fixed frequency with the same values of parameter *L*, the level of coherence in the general direction of wave propagation should be higher than in the transverse direction.

The results of field studies presented in this chapter (in particular, those presented in Fig. 3.1) indicate that the linear model of the wave field does not allow even a qualitatively correct description of the coherence dependence on the direction on scales close to the scale of dominant waves [48]. The data of the field measurements showed that within the range $1 \leq \Omega < 2$ the character of the anisotropy of the coherence determined according to (3.3) changes. At fixed distances at frequencies close to the frequency f_t of dominant waves, the coherence is higher f_t in transverse direction than in longitudinal direction $A_R > 1$. At higher frequencies above a certain transition frequency, the anisotropy of the coherence changes in the opposite direction, i.e. $A_R < 1$. This behavior of coherence anisotropy was observed in all our experiments under different weather conditions and at different distances between the wave gauges.

The estimation of anisotropy transition frequency f_t from one type to another was calculated based on the results of synchronous measurements of the quadratic coherence function in longitudinal and transverse directions. According to these data (41 series of measurements were made), obtained for different values of the dominant wave frequency, it was determined that the physical values of the transition frequency f_t were within the range from 0.29 to 0.78 Hz, or at a ratio 1:2.75. After the transition frequency by parametrization (3.4) was brought to a dimensionless form $\Omega_t = f_t / f_0$, the spread was reduced. The ratio of minimum to maximum Ω_t has entered the range of 1:2. The analysis of the measurement data showed that the mean value of the non-dimensional transition frequency $\overline{\Omega_t} = 1.4$ with a standard deviation of 0.3 [17].

The obtained estimation of the change frequency of the coherence anisotropy type practically coincides with the theoretical value of the dimensionless frequency $\Omega = \sqrt{2}$. For linear gravitational waves subject to the dispersion relation (3.1), the dimensionless frequency $\Omega = \sqrt{2}$ corresponds to a gravitational wave having a length $0.5\lambda_0$. Here λ_0 is the length of the dominant wave. The scale $0.5\lambda_0$ corresponds to the length of the second spatial harmonic of the dominant wave.

Thus, experimental studies carried out under natural conditions have shown that in the field of sea wind waves, the spatial scale equal to half of the dominant wavelength or corresponding frequency scale $\Omega = \sqrt{2}$ separates two regions radically different from each other in terms of anisotropy of coherence.

3.4 Comparison of Results of Field and Laboratory Experiments

This subsection compares the coherence loss rate in the direction of wave propagation as determined in laboratory and field experiments. For waves subject to the dispersion relation (3.1), the ratio

$$\frac{\omega^2 L}{g} = k\,L = \frac{2\pi\,L}{\lambda} \tag{3.5}$$

In the analysis of space-time characteristics of a wave field observed in different conditions it is expedient to use dimensionless parameter [37].

$$\varepsilon = L/\lambda \tag{3.6}$$

The parameter ε corresponds to the distance measured at wave lengths λ. Parameterization (3.6) is convenient both for generalization of measurement results and for theoretical analysis, as it allows to exclude the explicit frequency dependence and, accordingly, to reduce the number of variables.

By analogy with the correlation radius, let us introduce a parameter ε_x for the longitudinal quadratic coherence function where the coherence level drops below a certain set value. We will assume that this value is 0.5, i.e. $R_{\uparrow}^2(\varepsilon_{0.5}) = 0.5$.

Experimental studies on a fixed oceanographic platform have shown that values $\varepsilon_{0.5}$ vary widely from 0.17 to 0.64 [16]. The average value $\varepsilon_{0.5}$ is 0.32. Close estimates of the parameter $\varepsilon_{0.5}$ for longitudinal coherence are derived from data from the field experiment [44]. The values $\varepsilon_{0.5}$ calculated from the graphs of the quadratic coherence function given in this publication are in the range 0.13–0.32.

Similar studies conducted under laboratory conditions [26, 27] showed that values $\varepsilon_{0.5}$ differed markedly from those obtained under field conditions. Measurements were carried out in a ten meter tray with bases L equal to 8, 12, 16, ..., 64 cm, which is very close to the values of bases in the experiment described above on a stationary oceanographic platform. The air flow was created with the help of wind turbine generator, its speed was equal to 10, 12.5, 15 m/s. It was found that in a laboratory experiment significant coherence values were recorded at frequencies up to 8–9 Hz. Estimates of the parameter $\varepsilon_{0.5}$, defined from Fig. 9 in [26, 27], had values $\varepsilon_{0.5} = 1.1$ at air flow rate of 10 m/s and $\varepsilon_{0.5} = 2.1$ at air flow rate 15 m/s.

Among the available sources, only one other work devoted to the study of coherence in a large wind-water tunnel was found [29]. The illustration in this work (Fig. 7) made it possible to determine that under conditions of a large wind-water tunnel on small bases 3 and 10 cm the $\varepsilon_{0.5}$ are close to the level 0.65, i.e., lower than in the tray but higher than in the sea. The wind-water tunnel in which the measurements were made was 40 m long.

The field measurement data analyzed here covered a wide range of dominant wave sizes from 2 to 35 m, with the parameter $\varepsilon_{0.5}$ values not exceeding the 0.15–0.5 interval. In laboratory measurements, where it is difficult to create long waves, shorter wavelengths of 0.2–0.6 m were generated. The generalized results of calculations of estimates $\varepsilon_{0.5}$, on the data obtained in the field and laboratory experiments, are presented in Fig. 3.3.

When conducting experimental studies in trays it is difficult to study the characteristics of wind surface waves in the orthogonal direction to the wind velocity vector. Apparently, no data on laboratory studies of transverse coherence could be found among available publications.

Thus, the comparison of experimental data obtained under different conditions showed the difference between laboratory and sea waves of decimeter and meter ranges. This difference is that longitudinal coherence in short laboratory waves is preserved at distances of 1–2 wavelengths, while in sea waves disappears at distances of less than half the wavelength.

The possibility of using the results of laboratory research in the interpretation of sea-wave measurements should be investigated on a case-by-case basis. This fully applies to the study of spatial and temporal relationships in the wave field. A faster

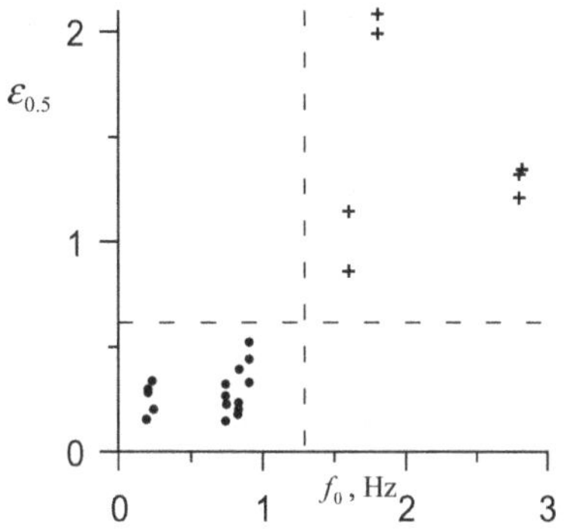

Fig. 3.3 Dependence of estimates $\varepsilon_{0.5}$ on the frequency of dominant waves f_0 in natural (•) and laboratory (+) experiments

loss of coherence in short-period sea waves may be caused by the orbital motions of the dominant wind waves and swell. The angular distribution of the waves in the trays is narrower than in the sea, and as will be shown below, as the angular distribution expands, the level of coherence decreases. In addition, there are other factors in the sea that lead to a decrease in the level of coherence. Such factors are turbulent fluctuations in surface current, variations in wind speed and direction, and the group structure of sea waves.

3.5 Modeling of the Quadratic Coherence Function

3.5.1 Angular Distribution Effect of Wave Energy

Let's consider a spatially homogeneous, stationary wave field, which is a superposition of noninteracting waves. Cross-spectrum of the wave displacements in two points connected by a vector $\vec{L}$, it is possible to write down as [37].

$$X(\omega, \vec{L}) = \int_0^{2\pi} \Psi(\omega, \alpha) \exp(i\,k\,L \cos\,\alpha) d \tag{3.7}$$

where $\Psi(\omega, \alpha)$ is the frequency angle spectrum; α is the angle between vectors $\vec{L}$ and $\vec{k}$. Let's represent the frequency-angle spectrum in the form of

$$\Psi(\omega, \alpha) = S(\omega)\Theta(\omega, \alpha) \tag{3.8}$$

where $\Theta(\omega, \alpha)$ is the spreading function that satisfy the condition of normalization

$$\int_{0}^{2\pi} \Theta(\omega, \alpha) d\alpha = 1 \tag{3.9}$$

If all waves are subject to the dispersion relation (3.1), then it follows from (3.7) and (3.8)

$$\mathrm{X}\left(\omega, \vec{L}\right) = \int S(\omega)\Theta(\omega, \alpha) \exp\left(i\frac{\omega^2 r_0}{g} \cos\alpha\right) d\alpha \tag{3.10}$$

or, considering (3.5) and (3.6)

$$\mathrm{X}(\omega, \varepsilon, \alpha_\Theta) = S(\omega) \int_{0}^{2\pi} \Theta(\omega, \alpha) \exp(i2\pi\ \varepsilon \cos\alpha) d\alpha \tag{3.11}$$

where the parameter α_Θ is determined by the angle between the vector $\vec{L}$ and the main direction of wave propagation. Thus, expression (3.2) can be given as follows

$$R^2(\varepsilon, \alpha, \alpha_\Theta) = \left| \int_{0}^{2\pi} \Theta(\omega, \alpha) \exp(i2\pi\ \varepsilon \cos\ \alpha)\ d\alpha \right|^2 \tag{3.12}$$

By using proven models of spreading functions, the expression (3.12) allows to construct theoretical estimates of the quadratic functions of longitudinal and transverse coherence for a linear wave field, as well as to estimate anisotropy of coherence. In order to compare the theoretical estimates with the experimental data, we use the most known models of the spreading function proposed in the works [6, 12, 28].

$$\Theta_1(\alpha) = \begin{cases} N_1 \cos^m \alpha; & |\alpha - \alpha_0| \le \frac{\pi}{2} \\ 0; & |\alpha - \alpha_0| > \frac{\pi}{2} \end{cases} \tag{3.13}$$

$$\Theta_2(\alpha) = N_2 \cos^{2s}\left(\frac{\alpha - \alpha_0}{2}\right) \tag{3.14}$$

$$\Theta_3(\alpha) = N_3 \mathrm{sech}^2(\beta(\alpha - \alpha_0)) \tag{3.15}$$

where N_i is the coefficients determined by the condition of normalization; α_0 is the main direction of wave propagation; m, s, β are dimensionless parameters determining the width of the angular distribution.

Parameters m, s,β are functions of dimensionless frequency Ω. In the region of main energy-bearing waves, the narrowest angular distribution is observed at the frequency of spectral peak, where $\Omega = 1$. As we move away from $\Omega = 1$ to both lower and higher frequencies, the angular distribution expands. In the model (3.14), parameter s depends on the stage of wave field development.

The spreading functions calculated for different values of dimensionless frequency of the model are presented in Fig. 3.4. Here the index i corresponds to the model index. When building the model (3.14), it was assumed that it corresponds to the stage of wave field development close to the state of a fully developed wave.

with the modeling results. To perform this procedure, we convert the spatial and temporal scales by presenting the values of the physical distance between the wave gauge in a dimensionless form (3.6), and use a dimensionless frequency (3.4). From the functions of longitudinal and transverse functions of coherence, thus parameterized, samples of coherence values at fixed values of non-dimensional frequency $\Omega = 1, \sqrt{2}, 2, 3$ were made and constructed as a function of the parameter ε (Fig. 3.5). Significant with 99.7% confidence probability is the level of the quadratic coherence function greater than 0.2. Parameterized empirical functions $R^2_{\uparrow}(\Omega, \varepsilon)$ and $R^2_{\uparrow}(\Omega, \varepsilon)$, are compared with theoretical functions calculated on the basis of Eq. (3.12) and models (3.13)–(3.15).

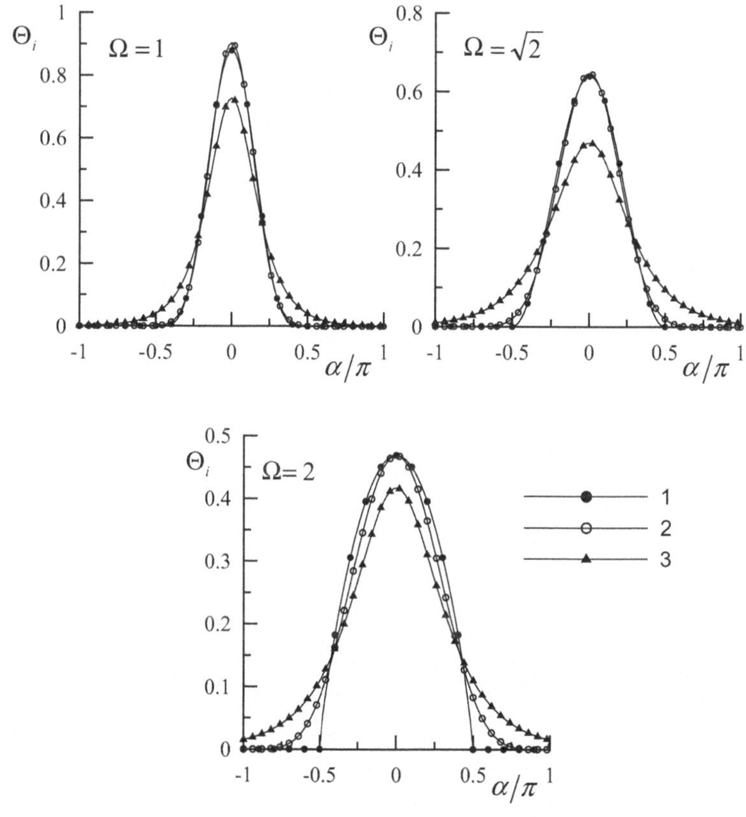

Fig. 3.4 Models of the spreading function. Curves 1–3 correspond to the models (3.13)–(3.15)

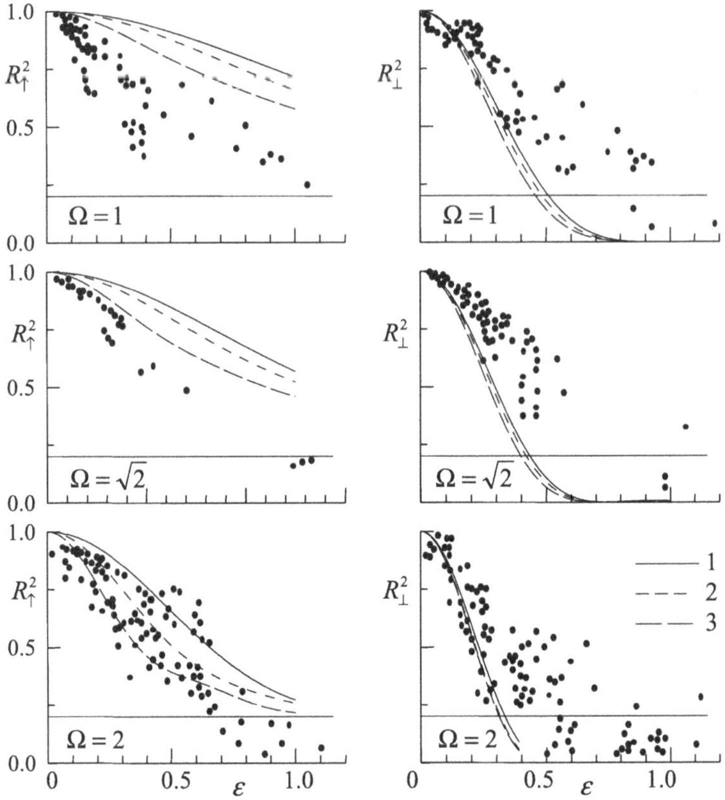

Fig. 3.5 Dependence of longitudinal $R_{\uparrow}^2(\Omega, \varepsilon)$ and transverse $R_{\perp}^2(\Omega, \varepsilon)$ functions of quadratic coherence on dimensionless distance ε. Curves 1–3 correspond to theoretical estimates of the quadratic coherence function based on models of the angular distribution function (3.13)–(3.15)

We will compare the quadratic coherence functions obtained in the experiments Presented in Fig. 3.5 graphs show that the coherence level, both longitudinally and transversely, decreases rapidly with growth ε. At values $\varepsilon = 0.5 - 1.0$ it drops below the statistical security level.

A decrease in the coherence level due only to the distribution of wave energy in the direction is illustrated by curves 1, 2 and 3 constructed for the three functions of angular distribution (3.13)–(3.15). In the above models, the following parameters were selected that determine the width of the angular distribution of the wave energy: at $\Omega = 1$ $m = 5$, $s = 10$, $\beta = 1.8$; at $\Omega = \sqrt{2}$, $m = 3.2$, $s = 7$, $\beta = 1.5$; at $\Omega = 2$ $m = 1.2$, $s = 3$, $\beta = 0.95$.

Behavior of theoretical functions $R_{\uparrow}^2(\Omega, \varepsilon)$ and $R_{\perp}^2(\Omega, \varepsilon)$ in Fig. 3.5 qualitatively corresponds to the behavior of the quadratic coherence functions constructed on the basis of measurement data: as the dimensionless distance ε increases, the calculated coherence values decrease rapidly. At the same time, it can be seen that there are systematic deviations of model curves from experimental estimates. These deviations are expressed in the fact that at $\Omega = 1$ and $\Omega = \sqrt{2}$, the longitudinal coherence values are higher than the measured values, while at the transverse coherence the calculated values are, on the contrary, lower than the measured values.

The deviations observed cannot be eliminated by selecting the angle function model or its parameters. By changing the width of the angular distribution of wave energy, the theoretical curves can be approximated to the experimental values for only one of the two directions (either longitudinal or transverse), but the correspondence for the other (orthogonal to it) is further reduced.

On experimental values of the quadratic coherence function $R^2_{\uparrow}(\Omega, \varepsilon)$ and $R^2_{\perp}(\Omega, \varepsilon)$ built polynomial regressions of the second order, whose coefficients are presented in Table 3.1. Here the parameter Δ means the standard deviation. Empirical functions of anisotropy of coherence $A_R(\Omega, \varepsilon)$ were constructed using these regression equations describing quadratic coherence functions. These experimental functions are then used to compare with the results of modeling spatial-temporal relationships in a wave field.

Figure 3.6 presents theoretical estimates of anisotropy of coherence built on three single-mode models of the angular distribution function (3.13)–(3.15). The use of different types of models leads to similar results. The principal differences between theoretical calculations and observations are shown: none of the models of the two-dimensional wave spectra allows to obtain the behavior of the coherence anisotropy function close to the experimental curve. At frequency $\Omega = 1$, the differences are not only quantitative but also qualitative, since all theoretical values lie in the region $A_R(\Omega = 1, \varepsilon) < 1$ and all experimental values lie in the region $A_R(\Omega = 1, \varepsilon) > 1$.

The isotropy of coherence i $A_R(\sqrt{2}, \varepsilon^*) = 1$ is observed at frequency $\Omega = \sqrt{2}$, while all theoretical values clearly indicate anisotropy. This fact also points to qualitative differences.

Table 3.1 Coefficients of regression $R^2(\dot{\varepsilon}) = a_0 + a_1\varepsilon + a_2(\varepsilon)^2$, quadratic function of longitudinal and transverse coherence $R^2_{\perp}(\Omega, \varepsilon)$ and $R^2_{\perp}(\Omega, \varepsilon)$

Quadratic function coherence	Polynomial coefficients				Coefficients of correlation, r
	a_0	a_1	a_2	Δ	
$R^2_{\uparrow}(\Omega = 1, \varepsilon)$	1.04	−1.50	0.72	0.09	0.87
$R^2_{\perp}(\Omega = 1, \varepsilon)$	1.03	−1.04	0.19	0.06	0.91
$R^2_{\uparrow}\left(\Omega = \sqrt{2}, \varepsilon\right)$	1.02	−0.92	0.08	0.03	0.98
$R^2_{\perp}\left(\Omega = \sqrt{2}, \varepsilon\right)$	1.02	−1.08	0.23	0.08	0.85
$R^2_{\uparrow}(\Omega = 2, \varepsilon)$	0.99	−0.92	0.23	0.10	0.83
$R^2_{\perp}(\Omega = 2, \varepsilon)$	1.03	−2.14	1.19	0.11	0.88

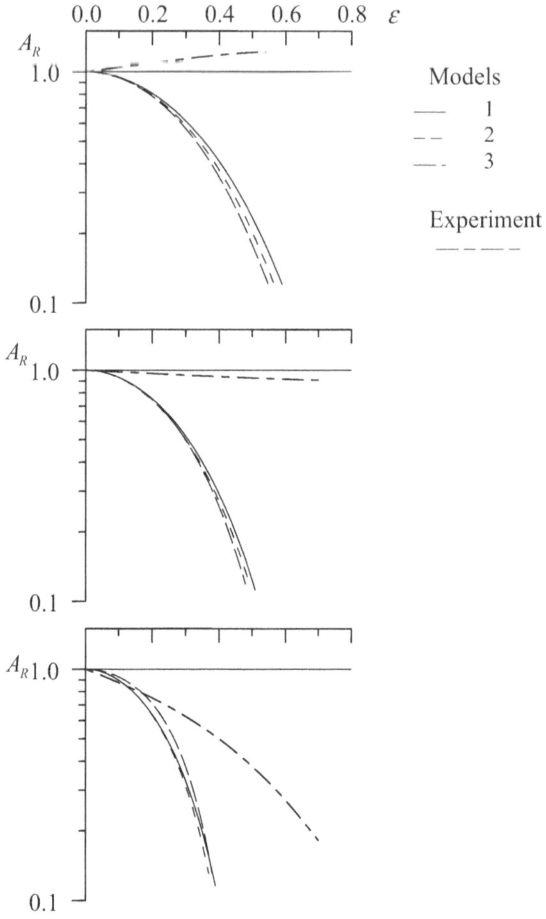

Fig. 3.6 Model and experimental dependencies of coherence anisotropy A_R on dimensionless distance ε. Curves 1–3 are constructed by models (3.13)–(3.15)

3.5.2 *Effect of the Presence of Longwave Harmonics*

Let us consider a wave field consisting of free waves subordinated to the dispersion relation for gravitational waves on deep water (3.1), and related components, which are harmonics of the dominant wave. The phase velocity of the harmonics is equal to the phase velocity of the dominant wave C_0. It follows from (3.1) that $C_0 = g/\omega_0$. The harmonics of the dominant wave propagate with its phase velocity and obey the dispersion relation

$$\omega = C_0 k \tag{3.16}$$

The cross-spectrum of the simulated wave field, consisting of free and coupled waves has the form

$$\mathrm{X}(\omega, \overrightarrow{L}) = \int_0^{2\pi} \Psi_f(\omega, \alpha) \exp\left(i \frac{\omega^2 L}{g} \cos\ \alpha\right) d\alpha + \int_0^{2\pi} \Psi_b(\omega, \alpha) \exp\left(i \frac{\omega\ L}{C_0} \cos\ \alpha\right) d\alpha \tag{3.17}$$

where indexes "*f*" and "*b*" denote free and bound components; the direction $\alpha = 0$ is chosen to coincide with the direction of the vector $\vec{L}$. In numerical analysis we will limit ourselves to estimating the influence of the second harmonic of the dominant wave.

We will assume that the full spectrum of the simulated wave field is the sum of free wave and harmonic spectra. The level of harmonics is determined by its fraction in the full spectrum

$$\gamma(\omega) = S_b(\omega) / S(\omega) \tag{3.18}$$

where $S(\omega) = S_b(\omega) + S_f(\omega)$

By substituting the cross-spectrum $\mathrm{X}(\omega, \overrightarrow{L})$ presented in the form (3.17) for expression (3.2), we obtain that the square coherence function can be described by expression:

$$R^2(\varepsilon) = \left\{ \int_0^{2\pi} \left[(1-\gamma)\Theta_f(\alpha) \cos(2\pi\, \varepsilon \cos\ \alpha) + \gamma\ \Theta_b(\alpha) \cos(\pi\, \varepsilon \cos\ \alpha)\right] d\alpha \right\}^2 + \left\{ \int_0^{2\pi} \left[(1-\gamma)\Theta_f(\alpha)\ \sin(2\pi\, \varepsilon \cos\ \alpha) + \gamma\, \Theta_b(\alpha) \sin(\pi\, \varepsilon \cos\ \alpha)\right] d\alpha \right\}^2. \tag{3.19}$$

where $\Theta_f(\alpha)$ and $\Theta_b(\alpha)$—angular distributions of free and bound components of the wave field respectively.

It has been shown above that in field measurements on the scale of the second harmonic show that the coherence level, defined in the longitudinal direction, decreases faster than it follows from calculations on known models of spreading functions. It can be assumed that the lower coherence values than those derived from the linear wave field calculations are the result of the presence of dominant wave harmonics.

Let's test the validity of that assumption. For this purpose, let us use the model (3.19). It is natural to assume that the spreading function of all harmonics repeats the spreading function of the dominant wave. The spreading function of the free waves expands with frequency, i.e., the spreading function of the harmonics already expands at frequency $\Omega = 2$ than the spreading function of the free waves. For quantitative estimations we will use the model of the spreading function in the form (3.15). According to [6], the change in the width of the angular distribution for this model is described by an expression

$$\beta = \begin{cases} 2.44\left(\frac{\Omega}{0.95}\right)^{-1.3}, & 0.95 \leq \Omega \leq 1.6 \\ 1.24, & 1.6 < \Omega \end{cases} \tag{3.20}$$

Using approximation (3.20), for the spreading functions (3.15) we set the following values of the parameter β: for the dominant wave and its harmonics $\beta = 2.28$; for free waves at a non-dimensional frequency $\Omega = 2$ we can accept $\beta = 1.24$. The results of the calculations are shown in Fig. 3.7.

In a complex multicomponent wave field, representing the superposition of free waves and harmonics, the level of coherence is influenced not only by the presence of waves of the same frequency, propagating with different phase velocities, but also a change in the angular distribution of wave energy, which becomes narrower with an increase in the proportion of harmonics. It follows from Eq. (3.19) that a decrease

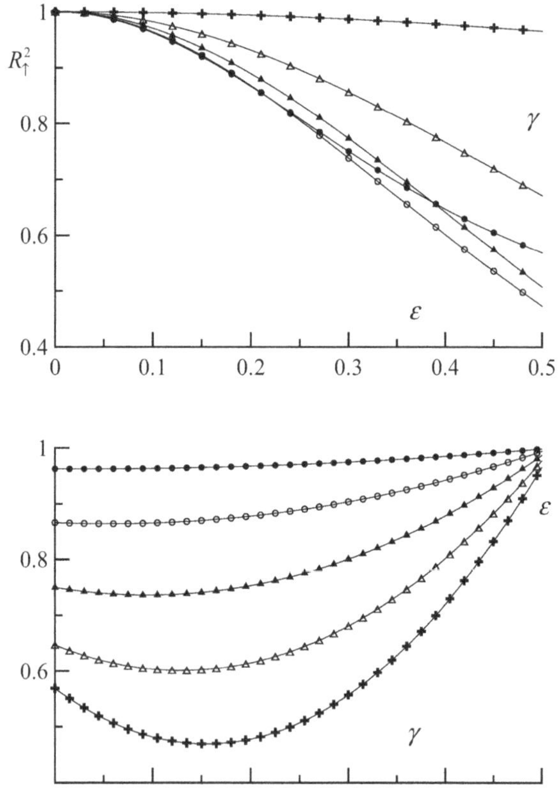

Fig. 3.7 Changes in the quadratic coherence function on the scale of the second harmonic at different harmonic levels

or increase in coherence when proportion of harmonics changes is also determined by how the width of the angular distribution of free and bound waves is related.

3.5.3 Group Structure Effect

A well-known feature of sea waves is their group structure. It is believed that its appearance is associated with nonlinear processes in the surface waves [1, 8], including the modulating instability of Benjamin-Feira [23].

The group and phase velocities of gravitational waves satisfying the dispersion relation (3.1) are equal to

$$C_G = d\omega/dk = \frac{1}{2}\sqrt{g/k} = \frac{1}{2}g/\omega \tag{3.21}$$

$$C_F = \omega/k = \sqrt{g/k} = g/\omega \tag{3.22}$$

i.e. the group speed is half as slow as the phase speed. As a result, by passing some distance, the wave profile is transformed, which leads to a decrease in the level of coherence in the direction of propagation.

Let us consider a nonlinear model of the wave field, assuming that the waves are one-dimensional. Let's assume that the waves propagate in the direction $\alpha = 0$. In this case, the spreading function can be presented in the form

$$\Theta(\alpha) = \delta(\alpha) \tag{3.23}$$

where δ is the Dirac Delta Function. The envelope of a weakly nonlinear wave train can be described by the nonlinear Schrödinger equation, whose established solution of the equation has a form [36]

$$G(x,t) = dn\left(m_g\left(k_0x - \frac{\omega_0}{2}t - \Phi_0\right), q\right) \tag{3.24}$$

where x—the spatial coordinate; dn—the elliptical function of Jacobi of the second kind with the module q. Parameters ρ_g and q determine the shape of the envelope and number of waves in the group, the parameter Φ_0 determines the position of the envelope at the initial moment.

Let us consider a case, when the one-dimensional carrier wave satisfies the dispersion relation (3.1) Modelled for the carrier wave with the envelope $G(x, t)$ changes of the surface level in the fixed point of space are given in Fig. 3.8. Calculations are carried out at three values of the parameter q equal to 0.1, 0.2 and 0.3.

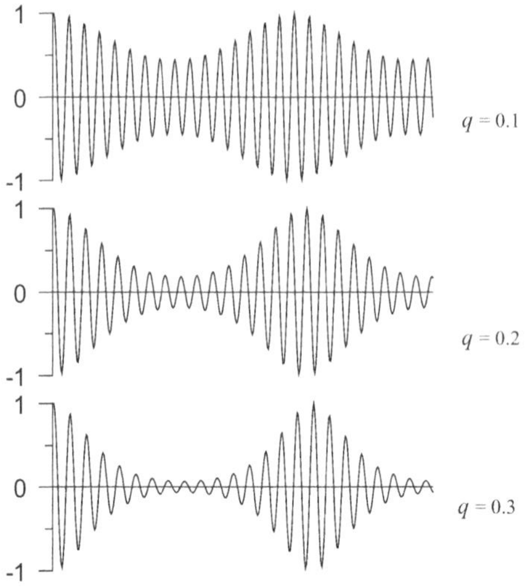

Fig. 3.8 Modeling of wavetrain with an envelope described by the expression (3.22)

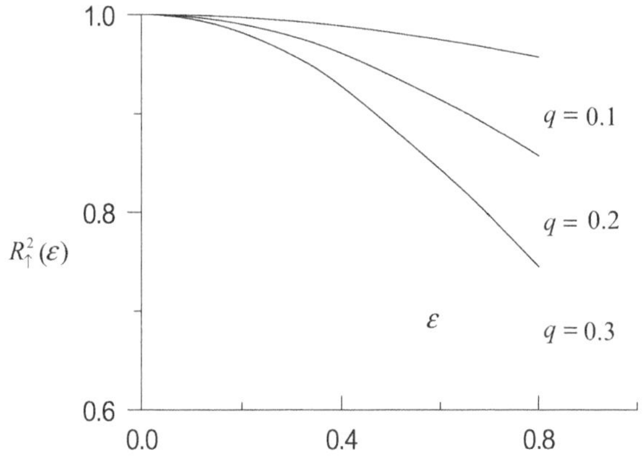

Fig. 3.9 Changes in the level of coherence of the carrier wave caused by the group structure

Changes in the quadratic coherence function $R_{\uparrow}^2(\varepsilon)$ are shown in Fig. 3.9. The group structure has a significant impact on the coherence level. With increasing magnitude q, the depth of the carrier wave modulation grows and the coherence level decreases.

Thus, taking into account the group structure of sea waves allows us to bring together the measured and model estimates of the longitudinal coherence function, because, as was shown above (Fig. 3.9), the decrease in coherence of the dominant components with the growth of the dimensionless distance ε in the real wave field is faster than it follows from the calculations of the known models of angular distribution functions (Fig. 3.5).

Numerical experiments were also conducted with other types of models describing the group structure of surface waves [48]. In these experiments, the correlation function of the surface elevation, rather than the envelope wave group, was defined. As in the above described numerical experiment with the model (3.24), it was assumed that the waves are one-dimensional. For approximation of the correlation function, two models were used

$$r_1(\tau, 0) = \exp(-\rho_1|\tau|)\cos(\omega_0\tau) \tag{3.25}$$

$$r_2(\tau, 0) = \exp(-\rho_2|\tau|)\cos(\widehat{\omega}\tau)\cos(\omega_0\tau) \tag{3.26}$$

where $\widehat{\omega}$—group frequency; ρ_1 и ρ_1—empirical parameters.

The approximation of the correlation function in the form (3.25) is valid only within one wave group, i.e. up to the first minimum of the envelope. This approximation corresponds to a wave spectrum with a single peak. The approximation in the form (3.26) allows to take into account the beats typical for the correlation functions of the wind wave and capillary wave, but it corresponds to the two-vertex spectrum with maxima at frequencies $\omega_0 \pm \widehat{\omega}$.

Taking into account the difference between phase and group velocities of gravitational waves, mutual correlation functions were simulated, on which the wave field cross-spectrum was built, and coherence losses corresponding to each model of correlation function were estimated with its help. The results of the numerical experiment confirm the conclusion that taking into account the group structure of sea waves makes it possible to converge the measured and model estimates of the longitudinal coherence function.

3.6 Calculation of Angular Distribution Functions

The measurement of the angular distribution function has been intensively carried out for several decades now [48]. However, interest in them has not waned, as evidenced by the large number of papers published in recent years (see, for example, [33, 41, 43]). This demonstrates the complexity of the problem and that it has not yet been definitively resolved.

As noted above, the calculation of the angular distribution function from the measurement data of an array of spaced sensors is based on the assumption of the linear nature of the sea wave field. This means that the cross-spectrum of the sea-wave field and related characteristics, such as the square coherence function and the phase spectrum, must be determined by two factors: the dispersion relation and the azimuthal distribution of the wave energy. If at frequencies above the frequency of the dominant waves there are effects due to the influence of the orbital motions of longer waves, the presence of harmonics that do not satisfy the dispersion equation for free waves, etc., then for the dominant waves this assumption looked quite correct.

The results of field studies and numerical simulations presented in this chapter show that there are good reasons to return to the question of the fairness of the assumptions made. There are a number of physical mechanisms that lead to changes in space-time relationships in the surface wave field and can significantly affect the results of calculations of the wave energy distribution along the directions.

It is natural to believe that a change in coherence caused by any physical mechanism individually should not lead to a decrease in coherence below the level observed in the experiment. However, the calculated coherence losses in the transverse direction only due to the azimuthal distribution of the wave energy turned out to be much greater than according to the natural measurements. This suggests that the actual angular distribution of dominant waves is narrower than it follows from parametrization of known models [2].

At present, information on the spatial spectra of sea waves is mainly obtained using wave buoys [25]. A way to determine the frequency-angle spectrum $\Psi(\omega, \alpha)$ from the data obtained with wave buoys was proposed in the paper [24]. To calculate the first five decomposition coefficients in the Fourier row of the frequency angle spectrum, the surface elevation and slope spectra in two orthogonal directions as well as their cross spectra are used. It is based on the assumption that the spatio-temporal characteristics of the wave field are determined by two factors: the dispersion equation and the azimuthal distribution of wave energy, i.e. the same assumption is used as the basis for the method for calculating the angular distribution on an array of sensors spaced out in space. However, as was shown above, taking into account only these factors is not enough to fully describe the space-time structure of the wave field. In particular, nonlinear effects in the surface wave field lead to the appearance of coupled components [11, 31]. This fact modifies the equations of binding spectra of elevation and slope of the surface, which, accordingly, affects the estimations of the values of decomposition coefficients in the Fourier function $\Psi(\omega, \alpha)$.

3.7 Conclusion

The series of experiments carried out on the fixed oceanographic platform of the Marine Hydrophysical Institute showed that spatio-temporal relationships in the sea-wave field cannot be described within the linear model. On the scale of the dominant waves, it was found that the coherence in the longitudinal direction (relative to the general direction of wave propagation) was decreasing faster than in the transverse direction. This effect may be related to the physical mechanisms that form the group structure of surface waves. Anisotropy of coherence (ratio of the quadratic coherence function defined in the transverse and longitudinal direction at a fixed distance) changes its character on the scale of the second spatial harmony of the dominant wave.

For short wave components (decimeter range) differences between laboratory and marine wind waves have been revealed. The difference lies in the fact that longitudinal coherence in laboratory waves is preserved at distances up to 1–2 wavelengths, and in marine conditions disappears at much shorter distances (about half the wavelength). The effect of rapid loss of coherence in the sea should lead to weakening of mechanisms of interwave interaction, carried out due to the implementation of the synchronism condition.

When measuring an array of wave gauges, the loss of coherence is determined by the angular distribution of wave energy and nonlinear physical mechanisms. If the coherence loss due to nonlinear physical mechanisms is not taken into account when determining the angular distribution function, the angular distribution function calculated from the multipoint measurements will be wider than the real one.

References

1. Arena F (2005) On non-linear very large sea wave groups. Ocean Eng 32(11–12):1311–1331
2. Babanin AV, Soloviev YP (1998) Variability of directional spectra of wind-generated waves, studied by means of wave staff arrays. Mar Freshw Res 49:89–101
3. Bakhanov VV, Demakova AA, Korinenko AE, Ryabkova MS, Titov VI (2018) Estimation of the wind wave spectra with centimeters-to-meter lengths by the sea surface images. Phys Oceanogr 3:177–190
4. Chung-Chu T, Bouchard R, Taft B (2004) Determination of pitch and roll angles from data buoys. Oceans '04 MTS/IEEE Techno-Ocean '04 (IEEE Cat. No. 04CH37600). https://doi.org/10.1109/oceans.2004.1405777
5. Danilyichev MV, Nikolaev AN, Kutuza BG (2009) The use of Kirchhoff method for practical calculations in microwave radiometry of the rough sea surface. Radiotehnika i elektronika 54(8):915–925
6. Donelan MA, Hamilton J, Hui WH (1985) Directional spectra of wind-generated waves. Philos Trans Roy Soc A315:509–562
7. Dugan JJ, Piotrowski CC (2003) Surface currents measured from a sequence of airborne camera images. In: Proceedings of the IEEE/OES seventh working conference on current measurement technology. https://doi.org/10.1109/ccm.2003.1194284
8. Fedele F, Tayfun MA (2009) On nonlinear wave groups and crest statistics. J Fluid Mech 620:221–239. https://doi.org/10.1017/S0022112008004424
9. Fujii S, Heron ML, Kim K, Lai J-W, Lee S-H, Wu X, Wu X, Wyatt LR, Yang W-C (2013) An overview of developments and applications of oceanographic radar networks in Asia and Oceania countries. Ocean Sci J 48(1):69–97
10. Graber HC, Terray EA, Donelan MA, Drennan WM, Van Leer JC, Peters DB (2000) ASIS—A New Air–Sea interaction spar buoy: design and performance at sea. J Atmos Oceanic Technol 17(5):708–720. https://doi.org/10.1175/1520-0426(2000)017%3c0708:aanasi%3e2.0
11. Hara T, Karachintsev AV (2003) Observation of Nonlinear Effects in Ocean Surface Wave Frequency Spectra. J Phys Oceanogr 33(2):422–430. https://doi.org/10.1175/1520-0485(2003)033%3c0422:ooneio%3e2.0.co;2
12. Hasselman DE, Dunckel M, Ewing JA (1980) Directional wave spectra observed during JONSWAP1973. J Phys Oceanogr 10:56–64
13. Ivonin DV, Telegin VA, Azarov AI, Ermoshkin AV, Bakhanov VV (2011) Possibility to measure velocity vector of surface currents by means of nautical radar with wide beamwidth antenna. Sovremennye problemy distantsionnogo zondirovaniya Zemli iz kosmosa 8(4):219–227. (In Russian)
14. Ivonin DV, Chernyshov PV, Kuklev SB, Myslenkov SA (2016) Preliminary comparisons of sea current velocity vector measurements by a nautical X-band radar and moored ADCP. Sovremennye problemy distantsionnogo zondirovaniya Zemli iz kosmosa 13(2):53–66

15. Khristoforov GN, Zapevalov AS, Smolov VE (1991) Longitudinal and transversal coherence in energy-carrier marine wind-waves. Izvestiya Akademii Nauk SSSR Fizika Atmosfery i Okeana 27(8):887–889
16. Khristoforov GN, Zapevalov AS, Smolov VE (1989) Measurements of collinear coherence in short-period sea wind-waves. Izvestiya Akademii Nauk SSSR Fizika Atmosfery i Okeana 25(6):636–643
17. Khristoforov GN, Zapevalov AS, Smolov VE (1995) Anisotropy of spatiotemporal correlations of sea wind-wave field Izvestiya Akademii Nauk Fizika Atmosfery i Okeana. 31(5):692–700
18. Krogstad HE, Trulsen K (2010) Interpretations and observations of ocean wave spectra. Ocean Dyn 60:973–991
19. Kudryavtsev V, Yurovskaya M, Chapron B, Collard F, Donlon C (2017) Sun glitter imagery of ocean surface waves: Part 1. Directional spectrum retrieval and validation, J Geophys Res Oceans 122(2):1369–1383
20. Kudryavtsev V, Yurovskaya M, Chapron B, Collard F, Donlon C (2017) Sun glitter imagery of surface waves. Part 2: waves transformation on ocean currents. J Geophys Res Oceans 122(2):1384–1399
21. Kuzmin AV, Goryachkin YA, Ermakov DM, Ermakov SA, Komarova NY, Kuznetsov AS, Repina IA, Sadovskii IN, Smirnov MT, Sharkov EA, Chuharev AM (2009) Marine hydrographic platform "Katsiveli" as a subsatellite test site in the Black sea. Earth Observation and Remote Sensing 1:31–44
22. Lake BM, Yuen HC (1978) A new model for nonlinear gravity waves. Part 1. J Fluid Mech 88:33–62
23. Lamont-Smith T, Fuchs J, Tulin MP (2003) Radar investigation of the structure of wind waves. J Oceanogr 59:49–63
24. Longuett-Higgins MS, Cartwrighte DE, Smith ND (1963) Observation of the directional spectrum of sea waves using the motions of the floating buoy. In: Proceeding of conference ocean wave spectra. Englewood Cliffs. NY Prentice Hall, pp 111–132
25. Mettlach T, Teng CC (2010) Concepts for an ideal ocean wave-measuring buoy. OCEANS 2010 MTS/IEEE SEATTLE. https://doi.org/10.1109/oceans.2010.5664525
26. Mitsuyasu H, Kuo YY, Musuda A (1979) On the dispersion relation of random gravity waves. Pt. 2. An experiment. J Fluid Mach 92:731–749
27. Mitsuyasu H, Kuo YY, Musuda A(1979) On the dispersion relation of random gravity waves. Pt. 2. An experiment. J Fluid Mach 92:731–749
28. Mitsuyasu H, Tasai F, Suhara T, Mizuno S, Ohkuso M, Honda I, Rikiishi K (1975) Observations of the directional spectrum of ocean waves using a cloverleaf buoy. J Phys Octanogr 5(4):750–758
29. Mollo-Christensen E, Ramamonjiarisoa A (1978) Modeling the presence of waves groups in a random wave field. J Geophys Res 83(C8):4117–4122
30. Panfilova M, Ryabkova M, Karaev V, Skiba E (2019) Retrieval of the statistical characteristics of wind waves from the width and shift of the doppler spectrum of the backscattered microwave signal at low incidence angles. IEEE Trans Geosci Remote Sensing. 1–7. https://doi.org/10.1109/tgrs.2019.2955546
31. Pelinovsky EN, Shurgalina EG (2016) Formation of freak waves in a soliton gas described by the modified Korteweg–de Vries equation. Doklady Phys 61(9):423–426
32. Pereira HPP, Violante-Carvalho N, Nogueira ICM, Babanin A, Liu Q, Ferreira de Pinho U, Nascimento F, Parente CE (2017) Wave observations from an array of directional buoys over the southern Brazilian coast. Ocean Dyn 67:1577–1591
33. Peureux C, Benetazzo A, Ardhuin F (2018) Note on the directional properties of meter-scale gravity waves. Ocean Sci 14(1):41–52. https://doi.org/10.5194/os-14-41-2018
34. Phillips OM (1981) The dispersion of short wavelets in the presence of a dominant long wave. J Fluid Mech 107:466–485
35. Plant WJ (2002) A stochastic, multiscale model of microwave backscatter from the ocean. J Geophys Res 107(C9):3120. https://doi.org/10.1029/2001JC000909

36. Pokazeev KV, Zapevalov AS, Pustovoytenko VV (2015) A nonlinear model of sea surface waves. Mosc Univ Phys Bull 70(3):213–215
37. Pokazeev KV, Zapevalov AS (2019) Calculation of phase velocities in the field of sea surface waves. Mosc Univ Phys Bull 74(4):413–418. https://doi.org/10.3103/S0027134919040143
38. Ramamonjiarisoa A, Coantic M (1976) Loi experimentale de dispersion de vagues produites par le vent sur une faible longueur d'action. C.R. Hehd Seances Acad Sci Ser. B. 282:111–113
39. Ramamonjiarisoa A, Giovanangeli JP (1978) Observations de la propagation des vagues engendrees par le vent au large. C.R. Hehd Seances Acad Sci Ser B 287:133–136
40. Rapizo H, Babanin AV, Schulz E, Hemer MA, Durrant TH (2015) Observation of wind-waves from a moored buoy in the Southern Ocean. Ocean Dyn 65(9):1275–1288
41. Simanesew AW, Krogstad HE, Trulsen K, Nieto Borge JC (2018) Bimodality of directional distributions in ocean wave spectra: a comparison of data-adaptive estimation techniques. J Atmos Oceanic Technol 35(2):365–384. https://doi.org/10.1175/jtech-d-17-0007.1
42. Wang DW, Hwang PA (2003) Higher fourier harmonics of the directional distribution of an equilibrium wave field under steady wind forcing. J Atmos Ocean Technol 20:217–227
43. Wyatt LR (2018) Measuring the ocean wave directional spectrum "First Five" with HF radar. Ocean Dyn 69(1):123–144. https://doi.org/10.1007/s10236-018-1235-8
44. Yefimov VV, Solov'yev P Yu, Khristoforov GN (1972) Observational determination of the phase velocities of spectral components of wind waves, Izv Atmos Oceanic Phys 8(4):435–446
45. Young IR (2006) Directional spectra of hurricane wind waves. J Geophys Res 111:C08020. https://doi.org/10.1029/2006JC003540
46. Yuen HC, Lake BM (1982) Nonlinear dynamics of deep-water gravity waves. Adv Appl Mech 22:67–229. https://doi.org/10.1016/s0065-2156(08)70066-8
47. Yurovskaya MV, Kudryavtsev VN, Stanichny SV (2019) Reconstruction of surface wave kinematic characteristics and bathymetry from Geoton-L1 multichannel optical images from "Resurs-P" satellite. Sovremennye problemy distantsionnogo zondirovaniya Zemli iz kosmosa. 16(2):218–226. (In Russian)
48. Zapevalov AS (1995) Estimation of the angular energy distribution function of the dominant sea waves. Izvestiya Akademii Nauk Fizika Atmosfery i Okeana 31(6):835–841
49. Zapevalov AS (2008) Statistical models of the sea surface in problems of acoustic and electromagnetic radiation scattering. Manuscript to claim the academic degree of doctor of physico-mathematical sciences on the speciality 04.00.22—geophysics. Marine Hydrophysical Institute of the National Academy of Sciences of Ukraine, Sebastopol, 2008
50. Zapevalov AS, Bol'shakov AN, Smolov VE (2004) Studies of the coherence level of sea surface waves. Izvestiya, Atmos Oceanic Phys 40(4):483–487
51. Zapevalov AS (2009) Bragg scattering of centimeter electromagnetic radiation from the sea surface: the effect of waves longer than Bragg components. Izvestiya Atmos Ocean Phys 45(2):253–261
52. Zapevalov AS, Bol'shakov AN, Smolov VE (2009) Studying the sea surface slopes using an array of wave gauge sensors. Oceanology 49(1):31–38

Chapter 4
Modeling the Shape of the Impulse Reflected from the Sea Surface

4.1 Introduction

At present, satellite altimetry is one of the main components of space monitoring of the World ocean. As an active method of remote sensing, satellite altimetry provides oceanographic information regardless of weather conditions and time of day. Its development has opened up new opportunities to study oceanological fields on a wide range of spatial and temporal scales and has led to significant progress in our knowledge of the dynamics of the World ocean.

In radio altimetry measurements, all sea level information along the spacecraft track is carried by the leading edge of the reflected radio impulse [7]. In addition to determining sea level, radio-sounding data determine two other sea surface parameters: the significant wave height [14, 27] and the wind speed [9, 27]. The possibility of remote determination of the senior statistical moments of the distribution of the excited sea surface elevations is being investigated [13]. The angle of slope of the leading edge allows to determine a significant wave height [7]. Changes in the distribution of sea surface elevations lead to changes in the shape of the reflected radio impulse [25, 36]. Changes in the shape of the reflected radio impulse result in errors in determination of sea surface level and significant wave height.

Most radio altimeters installed on satellites operate at wavelengths ranging from 2 to 6 cm. They have a narrow directional pattern with a width of no more than 3.4° (the exception being the altimeter mounted on the Envisat satellite with a width of 5.25°). The duration of the probing impulse varies widely from 1 to 100 μs. The impulse reflected from the sea surface undergoes a compression procedure, after which its duration is several several nanoseconds [8].

The main factor determining sea level measurement errors is the change in the state of its surface [16, 36]. Usually, three error groups are distinguished [12]. The first group includes errors due to deviation of sea surface elevation distributions from

A. Zapevalov et al., *Simulation of the Sea Surface for Remote Sensing*, Springer Oceanography,
https://doi.org/10.1007/978-3-030-58752-9_4

Gaussian distribution [2, 28]. The errors of the second group are caused by the fact that the reflection intensity of radio waves varies along the profile of the dominant surface wave, since the variance of sea surface slopes near the crest is higher than in the trough [21, 29]. The third group of errors is related to the pre-processing of recorded signals on board the satellite [16]. Here analyses the errors related to the first group.

The main model describing the distribution of sea surface elevations in remote sensing applications is the one based on the truncated Gram–Charlier series [13, 17];). Its advantage is a rather compact analytical form, and also that coefficients of a series are calculated on statistical moments of the studied parameter. The disadvantage is that in experiments, as a rule, statistical moments not older than the fourth order are defined. As a result, only the first five (including zero) members of the Gram–Charlier series can be used when constructing the probability density function. Models based on the truncated Gram–Charlier series incorrectly describe the distribution of surface elevations in the vicinity of large wave crest [35].

This Chapter analyses the influence of sea surface elevation distribution deviations from Gaussian distribution on the reflected signal shape of the satellite radio altimeter and also analyses sea surface level errors caused by non-linearity of wind waves. It also analyses the physical limitations of increasing the spatial resolution of spaceborne altimeter.

4.2 Brown's Model

When the sea surface is vertically probed, the signal recorded on the spacecraft is generated by a mirror reflection. This physical mechanism is the main one when generating a signal for incidence angles up to 20°–25°, at large incidence angles the main physical mechanism that generates a radio signal recorded on a satellite is the Bragg (resonance) scattering [1, 4]. The form of the reflected altimeter pulse is described by the Brown model representing a convolution of three functions [6]. The two functions depend on the characteristics of the altimeter mounted on the spacecraft and the measurement conditions (orbital altitude). The third function is determined by the properties of the sea surface, it describes the distribution of mirror reflection points along the wave profile. The dependence of the reflected radio impulse characteristics on the characteristics of the sea surface makes it possible to solve the reverse problem and to reconstruct these characteristics [20, 32].

Brown's model looks like

$$V(t) = \chi(t) * s(t) * q(t) \tag{4.1}$$

where $\chi(t)$ is the shape of the impulse reflected from a flat surface; $s(t)$ is the shape of the probing impulse; $q(t)$ is the function associated with the density of probabilities of the height of the mirror reflection points; symbol * means convolution; t is time. The Brown model was constructed under the following basic assumptions:

- The reflective surface contains a fairly large number of independent reflective elements;
- Surface elevation statistics are constant within the radar-lit surface;
- Reflection is a process without polarization effects;
- The normalized backscatter cross section is constant for the entire illuminated area, i.e. within the directivity diagram independent of the angle of incidence;
- The Doppler effect is negligible.

It should also be noted that mirroring takes place under the condition that the surface wavelength is much longer than the radio wavelength. Small scale ripple, which does not meet this condition, causes diffuse scattering, which leads to a reduction in power of quasi-mirror reflected radar signal [4]. In order to take into account the diffuse scattering Fresnel's coefficient is replaced by the effective reflection coefficient [31]. The effective reflection coefficient varies from 0.45 to 0.50 when the wind speed changes from 1.5 to 15 m/s [10].

Let's consider the case where the shape of the sounding radio impulse is Gaussian.

$$s_r(t) = \frac{1}{\sqrt{2\pi D_r}} \exp\left(-\frac{t^2}{2D_r}\right) \tag{4.2}$$

where the parameter D_r determines the width of the radio impulse. Let us assume that probing is done at zero angle of incidence and that the directional diagram of the altimeter antenna is described by the Gauss function. Within these assumptions, the shape of the impulse reflected from the flat surface is as follows [17]

$$\chi(t) = a \exp\left[-\frac{\ln 4c}{\cos^2(\theta/2)h} t\right] H(t) \tag{4.3}$$

where a is amplitude; c is speed of light; θ is beam width of the antenna, determined by the level of half power; h is orbit height of the satellite; $H(t)$ is Heaviside unit function.

If the shape of the sounding impulse $s(t)$ and the shape of the impulse reflected from the flat surface are described by $\chi(t)$ expressions (4.2) and (4.3), their convolution is described by the equation

$$I(t) = \chi(t) * s_r(t) = \frac{a}{2} \exp\left[\delta^2 \frac{D_r}{2} - \delta t\right] \left[1 - \operatorname{erf}\left(\left(\delta - \frac{t}{D_r}\right)\sqrt{\frac{D_r}{2}}\right)\right] \tag{4.4}$$

where $\delta = \frac{\ln 4c}{\cos^2(\theta/2)h}$; $\operatorname{erf}(x) = \frac{2}{\sqrt{\pi}} \int_0^x \exp(-y^2)\, dy$ is error function.

It follows from Eqs. (4.1) and (4.4) that the shape of the reflected impulse can be described by a convolution of two functions, where the first function is determined by the characteristics of the altimeter and the second function is determined by the characteristics of the sea surface.

$$V(t) = I(t) * q(t) \tag{4.5}$$

Model (4.1) assumes that mirror reflection points are evenly distributed along the wave profile, i.e., the function $q(t)$ is unequivocally determined by the density of probability of sea surface elevations $P(\eta)$. From where should

$$q(t) = \frac{d\eta}{dt} P(\eta(t)) \tag{4.6}$$

Parameters η and t are related by the ratio

$$t = \eta/(c/2) \tag{4.7}$$

Since the transform (4.7) is linear, the absolute values of skewness and kurtosis and the higher cumulant distributions $q(t)$ and $P(\eta)$ are equal. Decreasing sea surface level results in longer radio impulse altimeter return times. Positive values of the parameter η correspond to upward direction, therefore, odd statistical distribution moments $q(t)$ should have the opposite sign than sea surface elevation moments. In particular, the $q(t)$ skewness of the distribution should have the opposite sign than the skewness of the elevation of the sea surface $P(\eta)$.

When calculating the reflected altimeter impulse shape, we will take the values of the parameters θ_w, $\sqrt{D_r}$ and h, as in the work [17], take the corresponding parameters of altimeter installed on SEASAT-1: $\theta = 1.6°$, $\sqrt{D_r} = 1.327$ nsec and $h = 8 \times 10^5$ m. Replacing the values of the specified parameters with the values corresponding to modern altimeters does not introduce any fundamental changes.

4.3 Sea Surface Elevation Probability Density Function Models

The Brown model was originally constructed under the assumption that the distribution of sea surface elevations is described by the Gaussian distribution. At present, the main model of the probability density function of sea surface elevations used in the analysis of altimetric measurements is the model $P_{G-C}(\eta)$ based on the truncated Gram–Charlier series [13]. The model $P_{G-C}(\eta)$ describes a weakly nonlinear field of surface sea waves and allows take into account the effects leading to the deviation of the distribution of surface elevations from the Gaussian distribution [24].

Edgeworth's A-type form of the Gram–Charlier series is used to describe the wind wave field [18]

$$P_{G-C}(\eta) = \sum_{n=0}^{\infty} a_n\, H_n\left(\frac{\eta}{\sqrt{\overline{\eta^2}}}\right)\frac{1}{\sqrt{2\pi}}\exp\left(-\frac{1}{2}\frac{\eta^2}{\overline{\eta^2}}\right) \tag{4.8}$$

where H_n is orthogonal Hermit polynomials of order n; a_n is coefficients of the series (4.8); $\overline{\eta^2}$ is variance of surface elevations. It is assumed that the average value of the surface elevation $\overline{\eta} = 0$

The coefficients of the Gram–Charlier series are calculated on the basis of statistical moment estimates. When measuring waves in marine conditions, statistical moments are determined no higher than order 4. Therefore, the calculation of the density of probabilities of sea surface elevations is limited to the first members of the Gram–Charlier series. For calculations of forms of the reflected impulse several forms of model are used $P_{G-C}(\eta)$ [13, 17].

$$P^{(1)}_{G-C}(\eta) = \frac{\exp\left(-\frac{\eta^2}{2\overline{\eta^2}}\right)}{\sqrt{2\pi\,\overline{\eta^2}}}\left[1 + \frac{A}{6}H_3\left(\frac{\eta}{\sqrt{\overline{\eta^2}}}\right)\right] \tag{4.9}$$

$$P^{(2)}_{G-C}(\eta) = \frac{\exp\left(-\frac{\eta^2}{2\,\overline{\eta^2}}\right)}{\sqrt{2\pi\,\overline{\eta^2}}}\left[1 + \frac{A}{6}H_3\left(\frac{\eta}{\sqrt{\overline{\eta^2}}}\right) + \frac{E}{24}H_4\left(\frac{\eta}{\sqrt{\overline{\eta^2}}}\right)\right] \tag{4.10}$$

$$P^{(3)}_{G-C}(\eta) = \frac{\exp\left(-\frac{\tilde{\eta}^2}{2}\right)}{\sqrt{2\pi\,\overline{\eta^2}}}\left[1 + \frac{A}{6}H_3(\tilde{\eta}) + \frac{E}{24}H_4(\tilde{\eta}) + \frac{A^2}{72}H_6(\tilde{\eta})\right] \tag{4.11}$$

where A and E are the skewness and kurtosis of the statistical distribution of the sea surface elevations. The Hermite polynomials included in Eqs. (4.9)–(4.11) are described by the expressions

$$H_3(x) = x^3 - 3x \tag{4.12}$$

$$H_4(x) = x^4 - 6x^2 + 3 \tag{4.13}$$

$$H_6(x) = x^6 - 15x^4 + 45x^2 - 15 \tag{4.14}$$

The limitation of the distributions built on the basis of truncated Gram–Charlier series is that they allow to describe the probability density only in a limited range of the parameter under study. In particular, as a result of comparison of model calculations and data of sea surface waves measurements, it was found that the range where it is possible to apply the models (4.10) corresponds to the condition [35].

$$\left|\eta/\sqrt{\overline{\eta^2}}\right| < 3 \tag{4.15}$$

The functions of probability density built by the models $P^{(1)}_{G-C}(\eta)$ $(i = 1,2,3)$ are presented in Figs. 4.1 and 4.2.

At their construction it was taken into account that according to the measurements in field conditions skewness and kurtosis lie in the following ranges [19].

$$-0.05 < A < 0.4 \tag{4.16}$$

$$-0.4 < E < 0.4 \tag{4.17}$$

Since the skewness and the kurtosis of sea waves do not correlate, the calculations were performed for two combinations of these parameters corresponding to the maximum deviations from the Gaussian distribution. The values A and E were chosen to correspond to the maximum deviations from the Gaussian distribution.

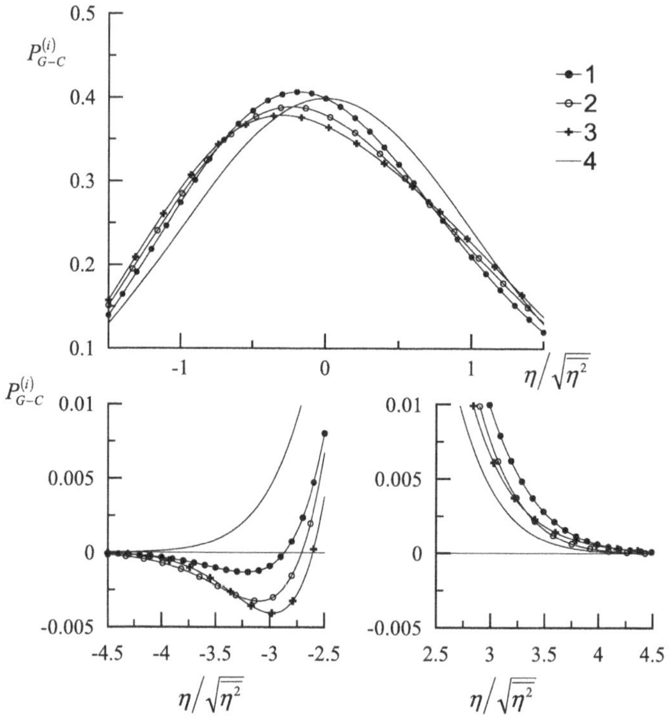

Fig. 4.1 Probability density function models of sea surface elevations. Curves 1–3 are models $P^{(i)}_{G-C}(\eta)$, model number i coincides with the number of the curve; curve 4 is the Gaussian distribution. The models $P^{(i)}_{G-C}(\eta)$ are constructed at $A = 0.4$, $E = -0.4$

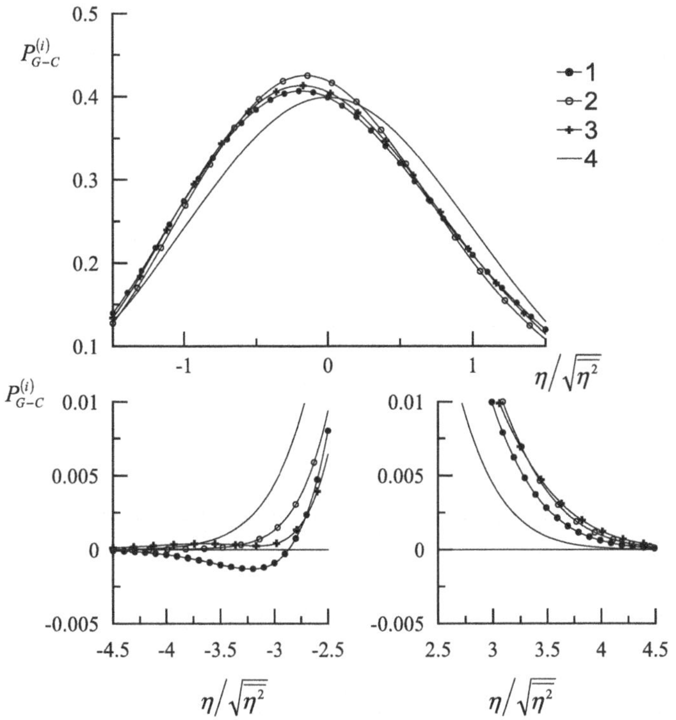

Fig. 4.2 Probability density function models of sea surface elevations. Curves 1–3 are models $P^{(i)}_{G-C}(\eta)$, model number i coincides with the number of the curve; curve 4 is the Gaussian distribution. The models $P^{(i)}_{G-C}(\eta)$ are constructed at $A = 0.4$, $E = 0.4$

If the values $\left|\eta/\sqrt{\overline{\eta^2}}\right|$ are large, all three functions $P^{(i)}_{G-C}(\eta)$ may have negative values. Let us consider what these distortions of the probability density function lead to when modeling the shape of the reflected impulse.

4.4 Reflected Impulse Shape Calculation

If the probability density $P(\eta)$ is described by the Gram–Charlier distribution in the form (4.11), then

$$q(t) = \frac{\exp\left(-\frac{t^2}{2D_s}\right)}{\sqrt{2\pi D_s}}\left[1 - \frac{A}{6}H_3\left(\frac{t}{\sqrt{D_s}}\right) + \frac{E}{24}H_4\left(\frac{t}{\sqrt{D_s}}\right) + \frac{A^2}{72}H_6\left(\frac{t}{\sqrt{D_s}}\right)\right] \tag{4.18}$$

where $D_s = (4/c^2)\overline{\eta^2}$.

For the case when functions $q(t)$ are described by the expression (4.18), the convolution [17]

$$q(t) * s_r(t = \frac{\exp\left(-\frac{t^2}{2D}\right)}{\sqrt{2\pi D}} \left[1 - \frac{\tilde{A}}{6} H_3\left(\frac{t}{\sqrt{D}}\right) + \frac{\tilde{E}}{24} H_4\left(\frac{t}{\sqrt{D_s}}\right) + \frac{\tilde{A}^2}{72} H_6\left(\frac{t}{\sqrt{D}}\right)\right] \tag{4.19}$$

where the parameters of Eq. (4.19) are given as

$$\begin{cases} D = D_s + D_r \\ \widetilde{A} = A_\eta (D_s/D)^{3/2} \\ \widetilde{E} = E_\eta (D_s/D)^2 \end{cases} \tag{4.20}$$

By integrating the expression on the right side term by term (4.19) we obtain the average form of the impulse reflected from the sea surface.

$$V(t) = \frac{a}{6} \exp\left[-d\left(\tau + \frac{d}{2}\right)\right] C(t) \tag{4.21}$$

where $\tau = \frac{t - t_0}{\sqrt{D}} - d$; $d = \frac{\ln 4}{\sin^2(\theta/2)} \frac{c}{h} \sqrt{D}$.

$$C = C_0 + \tilde{E}\, C_1 + \tilde{A}^2 C_2 \tag{4.22}$$

$$\begin{cases} C_0 = \frac{1}{\sqrt{2\pi}} \int\limits_{-\infty}^{\tau} [6 + A_\eta H_3(z + d)] \exp\left(-\frac{z^2}{2}\right) dz, \\ C_1 = \frac{1}{\sqrt{2\pi}} \int\limits_{-\infty}^{\tau} H_4(z + d) \exp\left(-\frac{z^2}{2}\right) dz, \\ C_2 = \frac{1}{\sqrt{2\pi}} \int\limits_{-\infty}^{\tau} H_6(z + d) \exp\left(-\frac{z^2}{2}\right) dz, \end{cases} \tag{4.23}$$

The convolution (4.19) allows the construction of a reflected impulse shape for the three models of sea surface elevation distributions (4.9)–(4.11). To switch from the model (4.11) to the models (4.9) and (4.10), one must assume that the coefficients C_1 and/or C_2 in (4.23) are zero. Built on the model (4.19), the forms of reflected radio impulse are presented in Fig. 4.3. The consequence of the distortions of the density function of surface elevation probabilities caused by the use of truncated Gram–Charlier series (as shown in Figs. 4.1 and 4.2) is not a physical effect. The effect is that in some situations negative values appear in the calculated form of the reflected impulse [25].

Comparison of impulse forms calculated by different models shows that the models $P_{G-C}^{(2)}(\eta)$ and $P_{G-C}^{(3)}(\eta)$ lead to almost the same results, but sharply differ

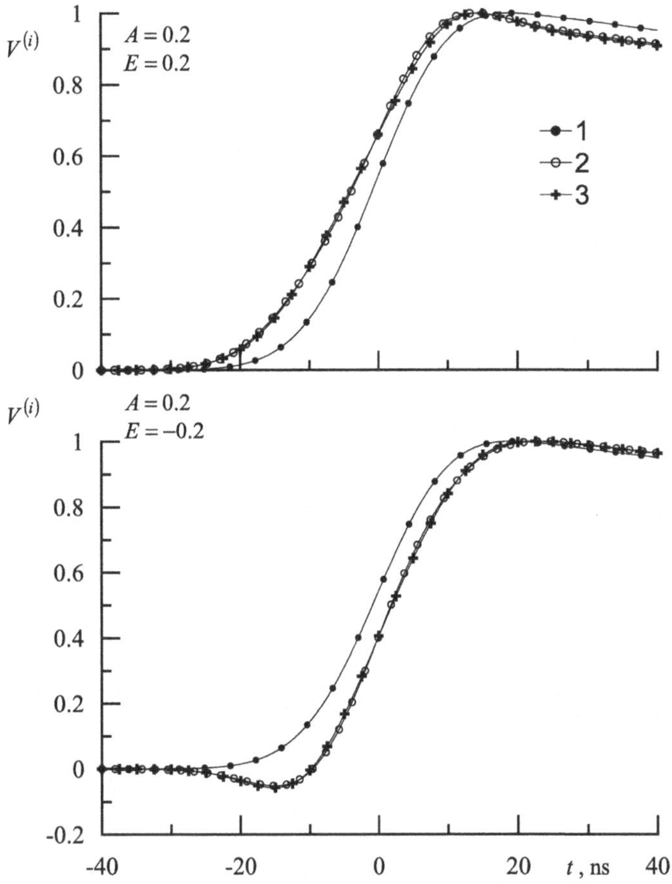

Fig. 4.3 The form of the reflected impulse $V^{(i)}$ calculated by three models of the sea surface elevations distribution (4.9)–(4.11). Index i corresponds to the number of the curve and the model $P^{(i)}_{G-C}(\eta)$ number

from the form obtained by the model $P^{(1)}_{G-C}(\eta)$. This points to the importance of including the sea surface elevation kurtosis in the sea level calculation procedure. The model $P^{(1)}_{G-C}(\eta)$ does not take into account the kurtosis.

As a rule, a significant wave height H_S equal to the average height of 1/3 of the highest waves is used as a criterion of surface wave energy. The significant wave height H_s is related to the variance of surface elevations by the ratio $H_S = 4\sqrt{\overline{\eta^2}}$[30]. Presented in Fig. 4.3 reflected impulse forms are obtained for $H_S = 5$ m.

It is possible to eliminate the disadvantages inherent in models of such kind $P^{(i)}_{G-C}(\eta)$ using a combined model, which in the range of small values of elevations

$\left(\left|\eta/\sqrt{\overline{\eta^2}}\right| < 3\right)$ is close to the Gram–Charlier model, outside of this range is close to the Gaussian distribution [26].

In general, a combined model based on the truncated Gram–Charlier series can be presented as [34].

$$P_C(\eta) = \frac{1}{\sqrt{2\pi\,\overline{\eta^2}}} \exp\left(-\frac{\tilde{\eta}^2}{2\,\overline{\eta^2}}\right)\left\{1 + F\left(\frac{\eta}{\sqrt{\overline{\eta^2}}}\right)\sum_{n=3}^{p} a_n\, H_n\left(\frac{\eta}{\sqrt{\overline{\eta^2}}}\right)\right\} \tag{4.24}$$

where F—is a filter, parameter p determines the number of members in a row.

The filter F should be close to one in the range $\left(\left|\eta/\sqrt{\overline{\eta^2}}\right| < 3\right)$ and tends to zero outside this range. As a filter, a two-parameter function is used

$$F\left(\frac{\eta}{\sqrt{\overline{\eta^2}}}\right) = \exp\left(-\left(\left|\frac{\eta}{\sqrt{\overline{\eta^2}}}\right|/d\right)^m\right) \tag{4.25}$$

where parameter d determines the range inside which the $F(\tilde{\eta}) \approx 1$ m, parameter determines the speed at which filter F tends to zero outside of this range. When simulating the distribution of sea surface characteristics, it is optimal to select the $m = 3.5$. $d = 3.4$ Note that the combined model may also have negative values if parameters m and d are incorrectly selected.

4.5 Errors in Sea Surface Level Determination

The accuracy of altimeter sea level measurements achieved to date significantly limits the range of oceanographic tasks that are carried out using data from spacecraft. The accuracy of altimeter sea level measurements is improved by refining the technical characteristics of the spaceborne altimeter (use of multi-frequency altimeters, improvement of ionosphere and tropospheric models, use of GPS precision positioning in space, etc.). The second way is to improve the information processing algorithms [3, 11, 22]. It is related to deeper understanding of the processes of reflected altimeter impulse formation as well as to new information on the structure and variability of the sea surface.

The level of the sea surface is determined by the transit time of the radio altimeter impulse from the satellite to the sea surface and back, which we denote as Δt. An error in the definition Δt by 1 ns leads to a sea level error of 15 cm. Let us quantify the displacements of the leading edge of the reflected radio impulse due to the deviation of the sea surface elevation distribution from the Gaussian distribution. The displacements will be calculated from the point corresponding to half the amplitude of the reflected radio impulse, the position of which on the time axis we denote as t_s.

The analysis will begin by investigating the effect of skewness on the accuracy of level determination. Let's consider how the position of the leading edge of the reflected impulse is changed when skewness and the constant significant wave height are changed. Taking into account that, as shown above, the forms of the reflected radio impulse, built on the models of distribution of elevations of the sea surface $P^{(2)}_{G-C}(\eta)$ and $P^{(3)}_{G-C}(\eta)$ coincide, at further analysis will be limited to the model $P^{(2)}_{G-C}(\eta)$.

The results of calculations presented in Fig. 4.4 show that as skewness increases, the leading edge of the reflected impulse shifts to the left, which corresponds to a decrease in the time of impulse passage to the sea surface and back, the calculated sea level increases. The magnitude of the displacement significantly depends on the kurtosis. The error in sea surface level determination caused by the skewness of sea surface elevations can be presented as

$$\delta_A = \frac{c}{2}[t_s(A, E, H_S) - t_s(A = 0, E, H_S)] \tag{4.26}$$

where $E = const$; $H_S = const$. Changes in the parameter δ_A with increasing skewness are shown in Fig. 4.5. Calculations were made for two values of significant height. It can be seen that the shift of the kurtosis in the range of negative values leads to a change of the sign δ_A. As the significant altitude increases, the error of sea level determination increases.

The analysis of sea level errors due to the deviation of the sea surface elevation distribution from the Gaussian distribution tends to focus on skewness changes [7, 12, 28]. The analysis presented here shows the necessity to take into account the kurtosis of waves simultaneously with their skewness.

4.6 Physical Limitations of Spatial Resolution of Spaceborne Radio Altimeter

An important characteristic that limits the range of geophysical problems solved using altimetric data, along with their accuracy, is spatial resolution. The use of synthetic aperture radar has significantly improved spatial resolution [5, 23]. When improving the spatial resolution, another physical factor that affects the accuracy of level detection must be considered. Remote sea level determination is done under the assumption that surface elevation statistics are constant within the radar area [6]. If the size of the irradiated surface area is comparable to the wave group length, the assumption of constancy ceases to be fulfilled, resulting in errors in restoring surface level and significant wavelengths. Let us estimate the errors arising in the case when

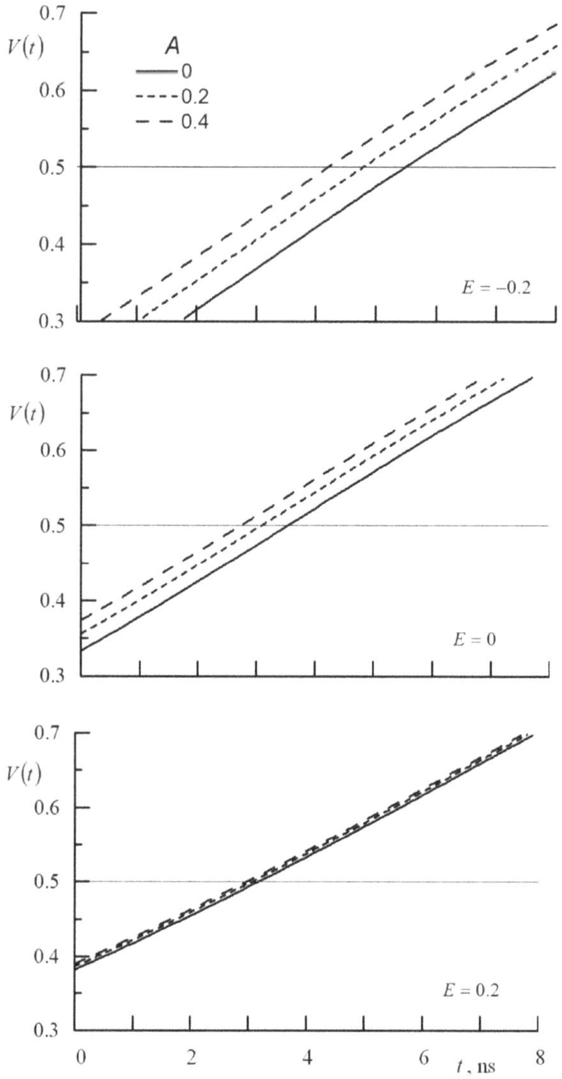

Fig. 4.4 Displacement of the leading edge of the reflected radio-altimeter impulse caused by deviations of the surface wave elevations distribution from the Gauss distribution. The significant wave height is equal to 5 m

the length of the local area L, for which the characteristics of the wave field are calculated, becomes comparable with the length of the wave group L_G.

Modeling the Group Structure of Surface Waves

The characteristic feature of the sea surface waves field is their group structure. Let us consider a one-dimension wave field, where t is time, x is the spatial coordinate. Let's represent the wave profile $\eta(x, t)$ as a product of the carrier wave and envelope. The wave profile is asymmetric, the distribution of elevations of the sea surface is quasi-Gaussian. Let us set the carrier wave in the form [33].

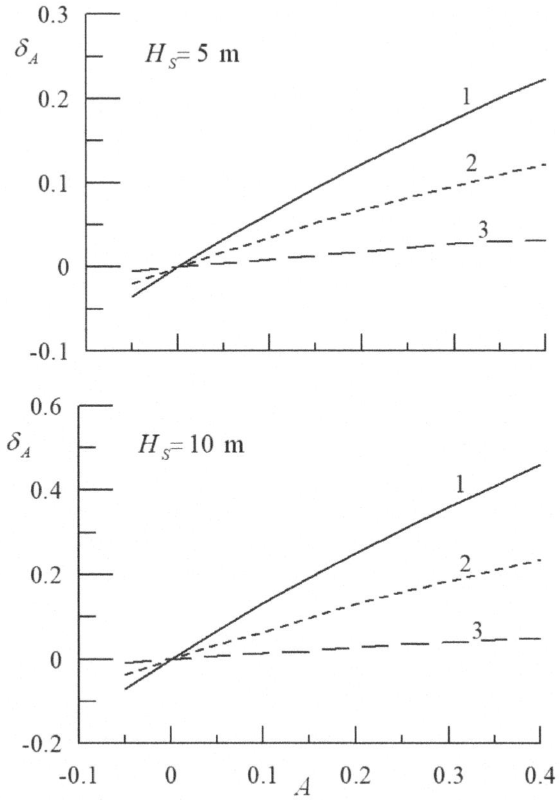

Fig. 4.5 Errors of altimetric sea surface level determination δ_A caused by skewness A deviation from zero; curve 1–3 corresponds to the kurt of −0.2, 0 and −0.2

$$w(x, t) = \exp\left[-\rho_0 \sin^2\left(\frac{k_0 x - \omega_0 t}{2}\right)\right], \tag{4.27}$$

where $k_0 = 2\pi/\lambda_0$ and ω_0—wave number and cyclic frequency of the carrier wave; λ_0—length of the carrier wave; the parameter ρ_0 determines the skewness of the carrier wave. Between themselves parameters k_0 and ω_0 connected dispersion relation for gravitational surface waves on deep water

$$\omega_0^2 = g k_0 \tag{4.28}$$

where g is gravitational acceleration.

The envelope groups of waves will be set in a similar (4.27) form

$$G(x, t) = \exp\left[-\rho_1 \sin^2\left(\frac{k_0 x - (\omega_0/2)\, t}{2\rho_2}\right)\right] \tag{4.29}$$

where parameters ρ_1 and ρ_2 determine the shape of the envelope and number of waves in the group respectively. In the final form, the wave profile has the form of

$$\eta(x,t) = \alpha\, G(x,t)\left(w(x,t) - \overline{w(x,t)}\right) \tag{4.30}$$

where α is the parameter that determines the wave heights; the line at the top means averaging.

The characteristics of the group structure of waves are the average number of waves in a group N_G and the group factor F_G [26].

$$N_G = \overline{\omega}/\overline{\Omega} \tag{4.31}$$

$$FG = 1.41\sigma_G/\overline{G} \tag{4.32}$$

where $\overline{\omega}$ is average frequency of wave spectrum; $\overline{\Omega}$ is average frequency of envelope spectrum; σ_G is standard deviation of envelope from its average value; $\overline{G}$ is average value of envelope. If to accept $\overline{\omega} = \omega_0$, then $\overline{\Omega} = \omega_0/(2\rho_2)$ from where should $N_G = 2\rho_2$.

The type of wave profiles constructed by the model described by Eqs. (4.27)–(4.30) at different values of the group factor *FG* is shown in Fig. 4.6. The physical meaning of the group-width factor is similar to the depth of modulation in radio engineering. (Фактор групповитости—везде *FG*) The parameter values *FG* determined under different conditions in field experiments usually range from 0.5 to 1. For the Black Sea, the average value of the group factor is 0.76 [26]. The presented wave profiles are normalized to significant wave heights H_S. On the abscissa axis, a dimensionless distance measured in the wavelengths λ_0 of the carrier wave is shown.

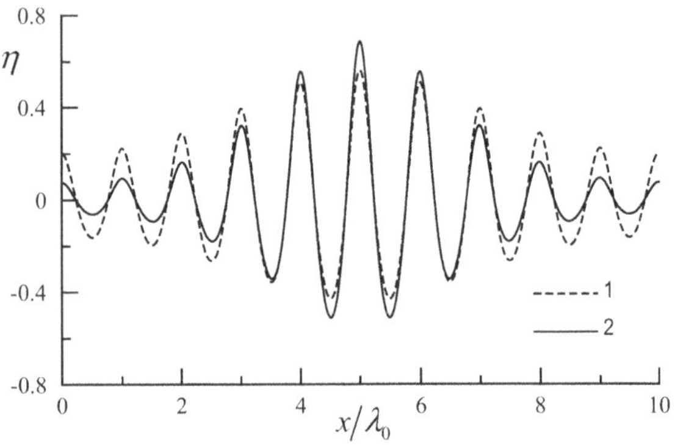

Fig. 4.6 Group structure of surface waves. Curve 1—$FG = 0.5$, Curve 2—$FG = 1$

4.7 Errors in the Determination of Sea Surface Characteristics Due to the Group Structure of Surface Waves

If the linear size of the area L for which the surface level is estimated is equal to or less than the wavelength of the group of waves, errors of two types may occur. First type: the average surface level in this area does not correspond to the level of the undisturbed surface. The second type: associated with errors in calculating the surface level caused by changes in statistical moments of surface elevation in this area.

Synthesized aperture altimeters make it possible to determine the level of sea surface areas with linear dimensions of several hundred meters [5], which is comparable to a wave group length. The length of the wave group L_G is determined by the number of waves in the group and their length, which depends on the conditions of wave field formation [15].

Depending on the position of the site relative to the envelope of waves, its average level may be above or below the level of the undisturbed surface. Let's enter the designation Δ_u—maximum displacement of the average level of the area up and Δ_d—down. Changes in parameters Δ_u and Δ_d depending on the ratio are L/L_G shown in the Fig. 4.7. In the case when the length of the section L is equal to the length of the wave group or is a multiple of it, then $\Delta_u = \Delta_d = 0$. Calculations are made for a significant wave height equal to 1 m. If the significant height increases, then the parameters Δ_u and Δ_d increase proportionately. If the group heights increase, the absolute values of the parameters Δ_u and Δ_d also increase.

A significant wave height H_S is one of the main characteristics of the sea surface that determines its condition. Estimation of accuracy of remotely determined parameters of the wave field is carried out by comparing *in situ* measurements of wave buoys [27]. At present, the error of parameter determination is H_S brought to an acceptable value for oceanology and is about 10% [22]. In estimating the accuracy of significant wave height measurements, instrumental error and errors created by an

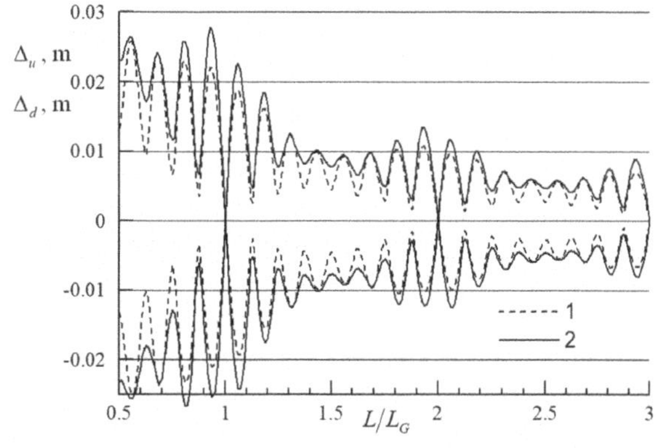

Fig. 4.7 Deviations of the middle level Δ_u and Δ_d section length L from the level of undisturbed surface: Curve 1—$F_G = 0.5$, Curve 2—$F_G = 1$

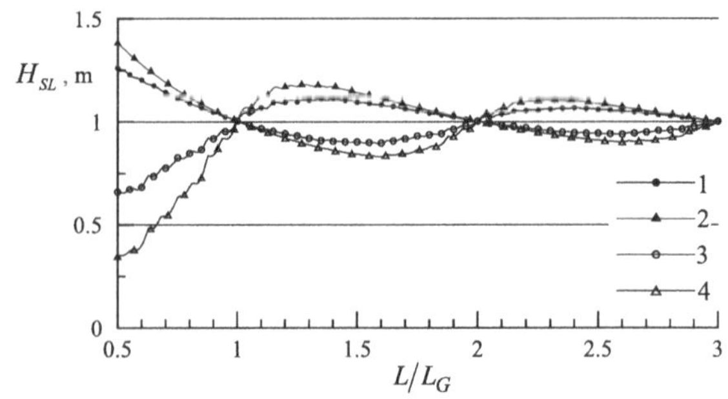

Fig. 4.8 Dependence of the limits of significant height change H_{SL} on the ratio L/L_G. Curve 1—maxH_{SL}, $FG = 0.5$; curve 2—maxH_{SL}, $FG = 1$; curve 3—minH_{SL}, $FG = 0.5$; curve 4—minH_{SL}, $FG = 1$

inaccurate coincidence in space and time of surface wave characteristics measurement from satellites and buoys are usually considered. If the linear dimensions of the section for which the significant wave height is calculated are comparable to the wavelength of the wave group, its local characteristic H_{SL} may differ from the average H_S.

The second statistical moment of the sea surface elevations, and consequently the significant wave heights H_{SL} calculated for a section of length *L*, depend on the ratio L/L_G. A section of sizes, which corresponds to the condition, $L < L_G$ may fall both on the region of large waves and on the region of small waves. The variance of the surface elevations calculated for this section can be either greater or smaller than the average value determined for the situation when $L \gg L_G$. Accordingly, depending on the position of the section relative to the group of waves, the values of significant heights H_{SL} are in some range from maxH_{SL} to minH_{SL}.

The limits in which the parameters H_{SL} change, when the ratio L/L_G changes, are shown in Fig. 4.8. The calculation was carried out at an average value of significant height equal to 1 m. To ensure that the error of determining the significant wave height of no more than 10% is guaranteed, it is necessary that the condition $L/L_G > 2$.

4.8 Conclusion

The potential for the practical use of data from altimeters installed on board spacecraft is limited by the accuracy of remote measurements of the sea surface topography achieved to date. This circumstance causes the great interest in the problem of reducing the errors of altimetry measurements.

At present, the improvement of altimeter information quality is due to the improvement of technical means of measuring the return time of the sounding radio impulse and improvement of algorithms of measurement data processing. The development of models for reflecting radio waves from the sea surface is largely due to the improvement of our knowledge about its structure and variability. Eliminating errors caused by changes in sea surface conditions will significantly improve the accuracy of sea level measurements.

Calculation of the form of the reflected impulse altimeter, by means of widely used models of probability distribution function of sea surface elevations, constructed on the basis of truncated Gram–Charlier series, has a number of drawbacks. The disadvantages are related to the limited range of elevations that describe these models. A clear manifestation of these distortions is the appearance of negative values in the calculated form of a reflected radio impulse.

Occurring as a result of nonlinearity of sea surface waves, deviations of statistical moments of its elevations from the values corresponding to Gaussian distribution, lead to change of the form of the reflected impulse altimeter. There is a shift of its leading edge, the position of which determines the time of signal passage from the satellite to the sea surface and back, i.e. there is an error of sea level determination. For different values of significant surface wave heights quantitative estimates of sea level recovery errors due to changes in the skewness and kurtosis of the surface elevation are obtained.

Improvements in the technical characteristics of space-based radars have significantly improved spatial resolution. When the spatial resolution of radio altimeters approaches the scales comparable to the wavelengths of a group of waves, there are physical limitations to the accuracy of determining the characteristics of the sea surface (level and significant wave height). Errors arise whose values depend on the ratio of the section length to the wavelength of the wavelength group. Level errors increase linearly with the growth of significant wavelengths. The group structure of sea waves is also a source of errors in determining significant wave heights. To ensure that the relative error of determining the significant wave height of no more than 10% is guaranteed, it is necessary to fulfill the condition: the length of the section for which it is calculated must be more than twice as long as the length of the wave group.

References

1. Barrick D (1968) Rough surface scattering based on the specular point theory. IEEE Trans Antennas Propag 16(4):449–454. https://doi.org/10.1109/tap.1968.1139220
2. Barrick DE, Lipa BJ (1985) Analysis and interpretation of altimeter sea echo. In: Saltzman B (ed) Satellite oceanic remote sensing. Advance in geophysics, vol 27, pp 61–100
3. Baskakov AI, Egorov VI (2008) Perspektivnyj vysokotochnyj sputnikovyj al'timetr (Advanced high-precision satellite altimeter). Sovremennye problemy distantsionnogo zondirovaniya Zemli iz kosmosa 5(1):113–119
4. Bass FG, Fuks IM (1979) Wave scattering by statistically rough surface. Pergamon, p 540
5. Bonekamp H, Montagner F, Santacesaria V, Nogueira Loddo C, Wannop S, Tomazic I, O'Carroll A, Kwiatkowska E, Scharroo R, Wilson H (2016) Core operational Sentinel-3 marine data product services as part of the copernicus space component. Ocean Sci 12(3):787–795. https://doi.org/10.5194/os-12-787-2016
6. Brown GS (1977) The average impulse response of a rough surface and its applications. IEEE Trans Antennas Propag AP-25:67–74
7. Callahan PS, Rodriguez E (2004) Retracking of Jason-1 data. Marine Geodesy 27:391–407. https://doi.org/10.1080/01490410490902098

8. Chelton DB, Walsh EJ, Macarthus JL (1989) Pulse compression and sea level tracking in satellite altimetry. J Atmospheric Oceanic Technol 6:407–438
9. Cheng Y, Xu Q, Liu Y, Lin H, Xiu P, Yin X, Zong H, Rong Z (2008) An analytical algorithm with a wave age factor for altimeter wind speed retrieval. Int J Remote Sens 29(19):5699–5716
10. Freilich MH, Vanhoff BA (2003) The relationship between winds, surface roughness, and radar backscatter at low incidence angles from TRMM precipitation radar measurements. J Atmospheric Oceanic Technol 20(4):549–562
11. Ginzburg AI, Kostianoy AG, Sheremet NA, Lebedev SA (2011) Satellite alti-metry application in the Black Sea. In: Coastal altimetry. Springer, Berlin Heidelberg, pp 367–388
12. Gómez-Enri J, Gommenginger CP, Challenor PG, Srokosz MA, Drinkwater MR (2006) ENVISAT radar altimeter tracker bias. Marine Geodesy 29:19–38
13. Gómez-Enri J, Gommenginger CP, Srokosz MA, Challenor PG (2007) Measuring global ocean wave skewness by retracking RA-2 Envisat waveforms. J Atmospheric Oceanic Technol 24:1102–1116
14. Grishechkin BY, Baskakov AI (2010) The potential of a space altimeter when measuring significant wave height. In: ISPRS TC VII symposium—100 years ISPRS, Vienna, Austria, July 5–7, 2010, IAPRS, vol XXXVIII, Part 7A, pp 49–52
15. Hasselmann K, Barnett TP, Bouws E, Carlson H, Cartwright DE, Enke K, Ewing JA, Gienapp H, Hasselmann DE, Kruseman P, Meerburg A, Mller P, Olbers DJ, Richter K, Sell W, Walden H (1973) Measurements of wind-wave growth and swell decay during the joint north sea wave project (JONSWAP). Ergnzungsheft zur Deutschen Hydrographischen Zeitschrift Reihe A(8)(12):1–95
16. Hausman J, Lotnicki V (2010) Sea state bias in radar altimetry revisited. Marine Geodesy 33(S1):336–347
17. Hayne GS (1980) Radar altimeter mean return waveforms from near-normal-incidence ocean surface scattering. IEEE Trans Antennas Propag AP-28:687–692
18. Huang NE, Long SR (1980) An experimental investigation of the surface elevation probability distribution and statistics of wind-generated waves. J Fluid Mech 101(1):179–200
19. Jha AK, Winterstein SR (2000) Nonlinear random ocean waves: prediction and comparison with data. In: Proceedings of 19th international offshore mechanics and arctic engineering symposium, ASME, 2000. Paper No. ETCE/OMAE 2000-6125
20. Karaev VYu, Kanevsky MB, Meshkov EM, Titov VI, Balandina GN (2008) Measurement of the variance of water surface slopes by a radar: verification of algorithms. Radiophys Quantum Electron 51(5):360–371. https://doi.org/10.1007/s11141-008-9042-6
21. Kumar R, Stammer D, Melville WK, Janssen P (2003) Electromagnetic bias estimates based on TOPEX, buoy, and wave model data. J Geophys Res 108(C11):3351. https://doi.org/10.1029/2002JC001525
22. Lavrova OYu, Kostyanoi AG, Lebedev SA, Mityagina MI, Ginzburg AI, Sheremet NA (2011) Kompleksnyi sputnikovyi monitoring morei Rossii (Integrated satellite monitoring of the Russia's seas). IKI RAN, Moscow, p 472
23. Le Roy Y, Deschaux-Beaume M, Mavrocordatos C, Borde F (2010) SRAL, a radar altimeter designed to measure several surface types. In: Proceedings of SPIE 7825, remote sensing of the ocean, sea ice, and large water regions 2010, p 782507, 18 October 2010, https://doi.org/10.1117/12.865060
24. Longuet-Higgins MS (1963) The effect of non-linearities on statistical distribution in the theory of sea waves. J Fluid Mech 17(3):459–480
25. Pokazeev KV, Zapevalov AS, Pustovoytenko VV (2013) The simulation of a radar altimeter return waveform. Mosc Univ Phys Bull 68(5):420–425. https://doi.org/10.3103/S0027134913050135
26. Presnukhin AV (2009) Group structure of wind waves in the Caspian sea. In: Lithodynamics of the bottom contact zone of the ocean. Proceedings of the international conference dedicated to the 100th anniversary of the birth of Professor V.V. Longinov, 14–17 September 2009, Moscow. GEOS, Moscow, pp 31–33

27. Queffeulou P (2004) Long-term validation of wave height measurements from altimeters. Marine Geodesy 27:495–510
28. Rodriguez E (1988) Altimetry for non-gaussian oceans: height biases and estimation of parameters. J Geoph Res 93(C11):14107–14120
29. Rodriguez E, Martin JM (1994) Estimation of the electromagnetic bias from retracked TOPEX data. J Geoph Res 99(C12):24971–24979
30. Stewart RH (2008) Introduction to physical oceanography. Texas A & M University, Prentice Hall, p 345. https://doi.org/10.1119/1.18716
31. Wu ST, Fung AK (1972) A noncoherent model for microwave emissions and backscattering from sea surface. J Geophys Res 77(30):5917–5929
32. Zapevalov AS (2018) Errors in radar measurements of the sea-surface large-scale slopes due to the sea-wave nonlinearity. Radiophys Quantum Electron 61(1):22–30. https://doi.org/10.1007/s11141-018-9866-7
33. Zapevalov AS (2017) Surface waves group structure modelling. Processes GeoMedia 2(11):497–502
34. Zapevalov AS, Pustovoitenko VV (2010) Modeling of the probability distribution function of sea surface slopes in problems of radio wave scattering. Radiophys Quantum Electron 53(2):100–110
35. Zapevalov AS, Bol'shakov AN, Smolov VE (2011) Simulating of the probability density of sea surface elevations using the Gram–Charlier series. Oceanology 51(3):406–413
36. Zapevalov AS (2012) Effect of skewness and kurtosis of sea-surface elevations on the accuracy of altimetry surface level measurements. Atmospheric Oceanic Phys 48(2):200–206

Chapter 5
Effect of Long Surface Waves on the Bragg Scattering of Microwave

5.1 Introduction

One of the major challenges facing modern oceanography, meteorology and climatology is the monitoring of surface wind fields at local, regional and global scales. Above the sea and ocean surface the problem is solved with the help of space technology means and methods—with the help of scatterometers, which, of course, do not directly measure the wind speed, but the power of the received radio signal, the magnitude of which determines the wind speed [17]. The level of sea surface roughness that determines the scattering and reflection of radio waves, along with changes in wind speed, is influenced by a large number of factors different in their physical nature [5]. The main elements of the sea surface that form the backscattered signal during remote sensing from spacecraft are short surface waves. The processes occurring in the boundary layers of the atmosphere and ocean, which are different in their physical nature, transform the field of short surface waves, making it possible to monitor these processes using radar equipment installed on spacecraft [3, 12, 17, 22]. The interpretation of remote sensing data requires a detailed study of the role of each of the processes affecting the sea surface in the formation of a radio signal recorded on a satellite.

The Bragg scattering creates surface waves whose length is comparable to that of a radio wave [24]. Long sea waves, compared to the probing radio wavelength, change the local slope of the surface (geometric effect) and modulate the energy of the Bragg components (dynamic effect). The interaction of radio waves with the sea surface is described in a two-scale model [27, 32].

The complexity of analysis of the interaction of radio waves with the sea surface is largely due to the lack of data on the structure and variability of the thin topographic structure of the surface itself [21]. In the absence of detailed information on its statistical characteristics, various models are often used. These are surface wave

A. Zapevalov et al., *Simulation of the Sea Surface for Remote Sensing*, Springer Oceanography,
https://doi.org/10.1007/978-3-030-58752-9_5

spectra models based on the solution of the kinetic equation [13] or numerical models of the sea surface topography [26]. The Gaussian model of the surface wave field is also used [6, 18, 25], which is constructed under the assumption that in a stationary, spatially homogeneous wave field, all components are independent and their phases are evenly distributed randomly.

The purpose of the studies, the results of which are presented in Chap. 5 (may write below, тогда лучше here), is to analyze the effect of long (as compared to the Bragg components) surface waves on the Bragg scattering of radio waves in the opposite direction. The analysis is based on numerical simulation using direct measurements of sea surface slope from the Marine Hydrophysical Institute oceanographic platform.

5.2 Bragg Scattering of Radio Waves by the Sea Surface

When the sea surface is probed at angles of incidence exceeding 20–25°, the interaction of radio waves with the sea surface is determined by the Bragg scattering mechanism. It is described in the framework of the small disturbances method [4]. If the Bragg (resonance) components of the surface wave field spread over a flat surface, the normalized cross section can be represented as

$$\sigma_{pp}^{0}(\mathrm{K}, \theta, \alpha) = 16\ K^{4} \left| G_{\ pp}(\theta) \right|^{2} \Xi_{r}\left(\vec{k}_{B}\right) \tag{5.1}$$

where θ is angle of incidence, α is azimuthal angle, $\left| G_{pp} \right|^{2}$ is the Bragg scattering geometric coefficient, which depends on the polarization of the emitted and received signal, the index pp indicates the type of polarization (the first index shows the polarization of the emitted signal, the second index shows the polarization of the received signal), Ξ is two-dimensional spectrum of sea surface elevation.

The Bragg scattering is created by waves propagating along the direction of sounding in forward and reverse directions, i.e. the direction of propagation of these waves is specified by conditions of

$$\varphi_{B} = \alpha, \quad \text{or} \quad \varphi_{B} = \alpha + \pi \tag{5.2}$$

Therefore the folded spectrum $\Xi_{r}\left(\vec{k}_{B}\right)$ is related to the directional wave spectrum $\Xi\left(\vec{k}_{B}\right)$ by [13]:

$$\Xi_{r}\left(\vec{k}_{B}\right) = \frac{1}{2}\left(\Xi\left(\vec{k}_{B}\right) + \Xi\left(-\vec{k}_{B}\right)\right) \tag{5.3}$$

The resonance condition connecting the wave numbers of the k_B sea surface waves and the wave numbers scattered in the reverse direction of the K radio waves is as follows

$$k_B = K2\sin\theta \tag{5.4}$$

There are several definitions of "long" and "short" waves. Here and elsewhere, we will consider as long those waves that satisfy the condition

$$\lambda \gg \lambda_B \tag{5.5}$$

where λ_B is the length of the Bragg wave. The presence of long waves leads to the fact that the Bragg components spread over a curvilinear surface [23], and to changes in the local angle of incidence of radio waves. As a result, the value of the geometric coefficient changes $\left|G_{pp}(\theta)\right|^2 \to \left|G_{pp}(\theta - \beta_\uparrow)\right|^2$. The Bragg wave component becomes a function of three variables $k_B = k_B\left(K, \theta, \beta_\uparrow\right)$. The expression describing the local normalized radar cross section of the sea surface in the presence of long waves takes the form

$$\sigma^0_{pp}(\mathrm{K}, \theta, \alpha, \beta_\uparrow) = 16\ K^4 \left|G_{pp}(\theta - \beta_\uparrow)\right|^2 \Xi_r\left(\vec{k}_B\left(K, \theta - \beta_\uparrow\right)\right) \tag{5.6}$$

where $\beta_\uparrow$ is the angle of long-wave tilt in the direction of the fall of radio.

Let us assume that the size of the radar spot on the sea surface is much larger than the size of long waves. This condition is generally met when sensing from spacecraft. The effect of the long waves on the value of the signal scattered backwards can be taken into account by averaging expression (5.6) over the whole range of angles of the sea surface $\beta_\uparrow$ created by them. Averaging should be carried out with a weight that determines the probability of the appearance of some or other angles $\beta_\uparrow$. Then [27],

$$\begin{aligned}\sigma^L_{pp} &= \int \sigma^0_{pp}(\theta - \beta_\uparrow)\, P(\beta_\uparrow)\, \mathrm{d}\beta \\ &= \int 16\ K^4 \left|G_{pp}(\theta - \beta_\uparrow)\right|^2 \Xi_r\left(\vec{k}_B\left(K, \theta - \beta_\uparrow\right)\right) P(\beta_\uparrow)\, \mathrm{d}\beta_\uparrow\end{aligned} \tag{5.7}$$

where $P(\beta_\uparrow)$ is the probability density function of angles $\beta_\uparrow$.

To carry out the averaging (5.7), we transform the elevation spectrum of the sea surface Ξ, presenting it as an explicit function of the angle of incidence. Let us passing on from the spectrum of wave vectors Ξ to the spectrum of wave numbers and directions $\chi(k, \varphi)$,

$$\Xi(k_x, k_y)\frac{\partial\left(k_x, k_y\right)}{\partial\left(k, \varphi\right)} = \chi(k, \varphi) \tag{5.8}$$

where Jacobian is $\frac{\partial\left(k_x, k_y\right)}{\partial\left(k, \varphi\right)} = k$. The spectrum $\chi(k, \varphi)$ will be appear in the form

$$\chi(k, \varphi) = S(k)\,\Theta(k, \varphi) \tag{5.9}$$

where $S(k)$ is spectrum of wave numbers; $\Theta(k, \varphi)$ is spreading function which describes angular distribution of wave energy. Spreading function meets the condition of normalization

$$\int_{-\pi}^{\pi} \Theta(k, \varphi)\, d\varphi = 1 \tag{5.10}$$

Let's replace the variables in the spectrum $S(k)$

$$S(k)\frac{dk}{d\theta} = \tilde{S}(K2\sin\theta) \tag{5.11}$$

where $\frac{dk}{d\theta} = K2\cos\theta$. In the final form for the backscatter cross section of the sea surface, we get

$$\sigma_{pp}^{0} = 2K^2 \left| G_{pp}(\theta) \right|^2 \frac{\tilde{S}(K2\sin\theta)}{\sin\theta\cos\theta}\, \Theta(k_B, \varphi_B) \tag{5.12}$$

The resulting expression allows to estimate changes of σ_{pp}^{0} caused by the presence of long waves on the sea surface.

We will analyze two situations where probing is done in the centimetre and millimetre wavelength ranges. In the first situation, we will consider long waves whose length λ_1 is more than 10 cm, in the second situation, waves are long if their length λ_2 is more than 2 cm. Experimental studies of senior statistical moments of slopes (skewness and kurtosis) created by surface waves in the range of wavelengths, the lower boundary of which is the spatial scale λ_1 or λ_2, previously, have not been conducted.

5.3 Statistical Characteristics of Sea Surface Slopes

Applications involving the scattering of electromagnetic waves on the sea surface typically use statistical slope moments obtained from optical images of the sea surface in work [9]. When investigating the effect of long waves on Bragg radio wave scattering, these data cannot be used. Mirror reflection creates waves much longer than the optical wavelength. The shortest surface wavelength is determined by the process of viscous dissipation and is equal to $\lambda_{dis} \approx 0.001$ m [1]. Obviously, λ_{dis} much longer than the optical wavelength. To analyze the effects created by long waves, it is necessary to use statistical moments of slopes created by waves whose length exceeds a certain length λ. The wavelength λ depends on the length of the probing radio wave and is determined by the condition (5.5).

To calculate the statistical moments of the sea surface slopes created by the waves of a given range, we will use data from direct measurements made on the oceanographic platform of the Marine Hydrophysical Institute. The oceanographic platform has been installed on the Black Sea near the Southern coast of Crimea. The depth of the sea at the place where the platform was installed is 30 m. For the Black Sea, this depth corresponds to the "deep water" condition, i.e. the wave field distortions caused by the bottom influence can be neglected [30].

The sea surface slopes on the oceanographic platform were measured using a two-dimensional laser slopemeter. The principle of operation of the device is based on measuring the laser beam deviation angles when passing the excited water-air interface from under water. The recorded angle of deviation of the laser beam from the vertical is determined by the local slope of the sea surface at the point where it is crossed by the beam. The measurement resolution was 0.02 s, the laser spot diameter on the undisturbed surface was about 1 mm. Measuring equipment and conditions of measurements are described in the works [29, 30].

At the specified measurement discreteness, sea surface slope records obtained with a laser gradiometer contain information on sea surface slopes in the frequency band up to 25 Hz. To eliminate the effects of high-frequency components, the original slope records were skipped through a low-frequency filter with a filter cut-off frequency of 4 Hz. Waves with frequencies below 4 Hz are those in the gravitational range that satisfy the dispersion relation of

$$\omega^2 = g\,k \tag{5.13}$$

where ω is circular frequency; g is gravitational acceleration. It follows from the dispersion relation that the frequency of 4 Hz corresponds to a wave with a length equal to 10 cm.

In the analysis of sea surface slope it is common to distinguish two components oriented along ξ_u and across ξ_c the wind direction. Here and further the index u corresponds to the upwind component of the slope, the index c corresponds to the crosswind component. The absence of an index means that the expression is true for both components of the slope. Let's enter the notation $\overline{\xi^2}$ is dispersion, A is skewness, E is kurtosis.

Statistical moments of slopes, in upwind and crosswind directions are presented in Fig. 5.1.

The skewness and kurtosis of the sea surface slope components created by waves longer than 10 cm indicate a quasi-Gaussian nature of their distributions. They lie in the same ranges as the statistical moments of slopes created by all waves present on the surface [7, 9]. The average value of skewness of the crosswind component of the slopes is close to zero, but in some situations there are significant deviations from zero. The skewness of the longitudinal component increases with wind speed. The explicit dependence of the kurtosis on wind speed is not observed.

It should be noted that in field experiments, the angle of inclination β of the sea surface is usually not determined. Determine its slope ξ, which is the first spatial derivative of surface elevation [7, 10, 21]. These parameters are related to each other

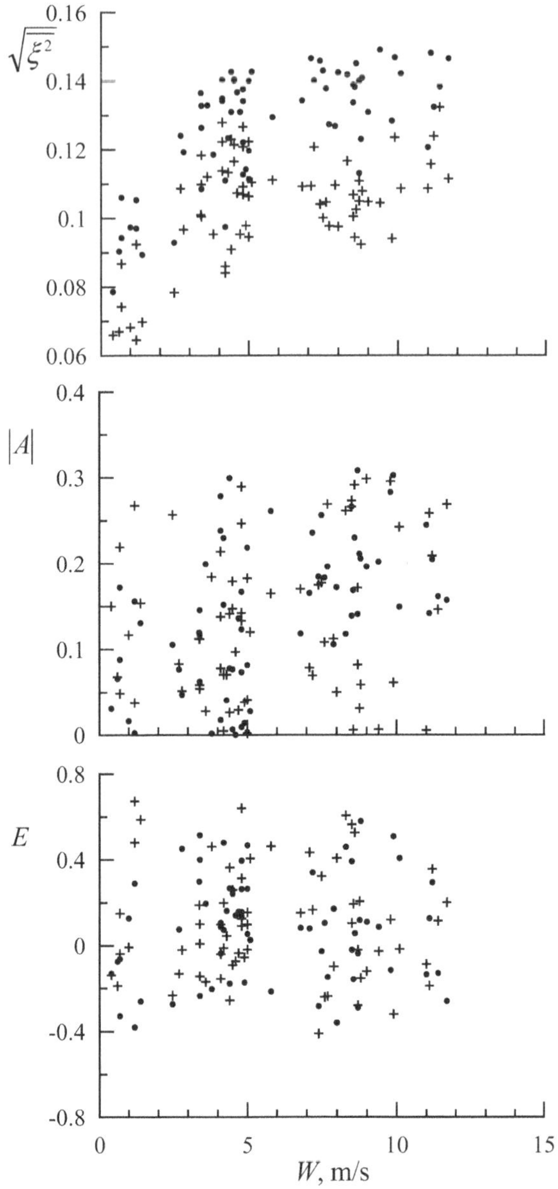

Fig. 5.1 Statistical moments of wind and crosswind slope components, point is upwind component; cross is crosswind component

by the formula

$$\beta = \arctan \xi \tag{5.14}$$

Since the dip angles created by surface waves are not large, it is generally assumed that the parameters β and ξ are numerically equal. Using this assumption introduces an error to the scattering model. The statistical moments of slopes and angles are shown in Fig. 5.2 [31].

According to measurements on the oceanographic platform for the same direction, the relationship between statistical moments of the same order of slope β and slope angles ξ is well described by linear regression equations $Y = ax + b$ [31]. Numerical

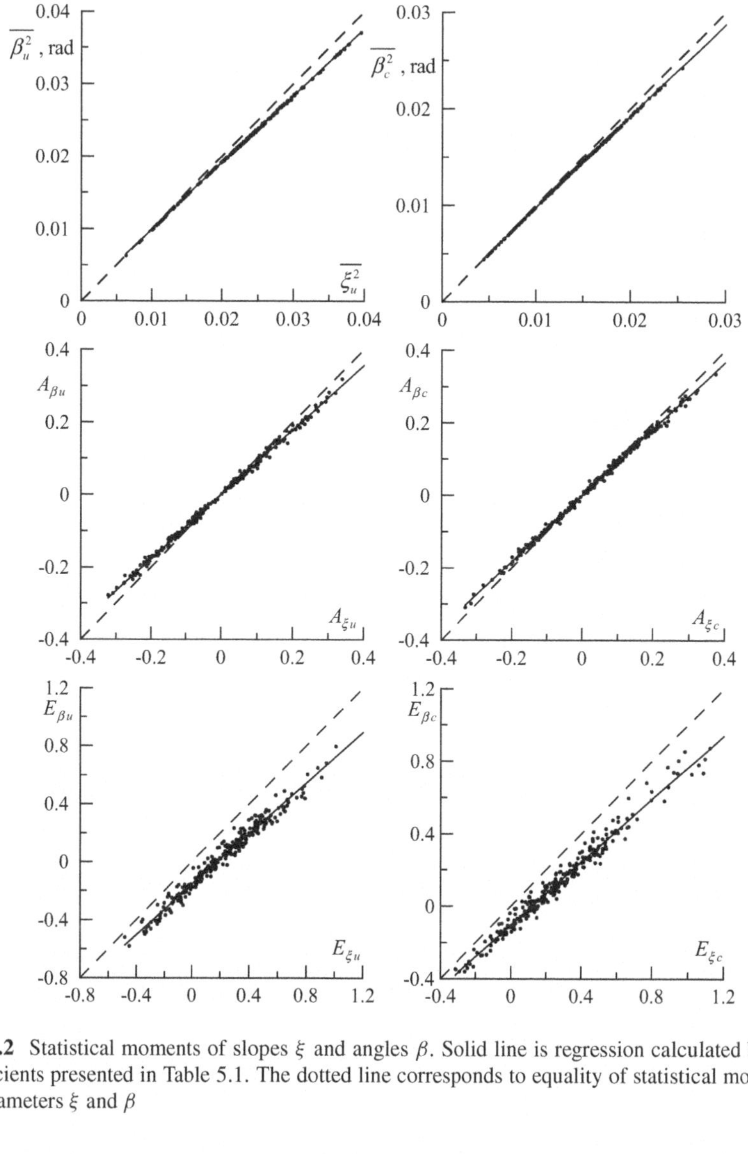

Fig. 5.2 Statistical moments of slopes ξ and angles β. Solid line is regression calculated by the coefficients presented in Table 5.1. The dotted line corresponds to equality of statistical moments of parameters ξ and β

Table 5.1 Coefficients of linear regression equations connecting statistical moments of slope angles and sea surface slopes

	Upwind component		
	a	b	δ
$\sqrt{\overline{\beta^2}}$	0.924	0	0.0001
A	0.889	0	0.0077
E	0.872	−0.155	0.040
	Crosswind component		
	a	b	δ
$\sqrt{\overline{\beta^2}}$	0.946	0	0.0001
A	0.916	0	0.007
E	0.869	−0.105	0.038

values of coefficients of this equation, and also values of standard deviation δ, are shown in Table 5.1.

The difference between the statistical moments of the parameters β and ξ increases with their order. The absolute skewness value for slope angles is smaller than for slopes, i.e. the distribution of the parameter β is more symmetrical than the distribution ξ. We can also state that the peaks of the probability density functions of the parameter are ξ sharper than those of β.

5.4 Sounding of the Sea Surface in the Centimeter Range of Radio Wave

5.4.1 Numerical Estimates of Changes in the Length of the Bragg Component

The estimates $\sqrt{\overline{\xi^2}}$ obtained in the experiments allow us to determine within what limits there is a change in the length of the Bragg component as a result of changes in the local angle of incidence caused by long waves. When spread over an inclined surface, created by long waves, the correlation between the length of the Bragg wave λ_B and the length of the radio wave Λ is

$$\lambda_B = \frac{\Lambda}{2\sin(\theta \pm \beta_\uparrow)} \tag{5.15}$$

The minus sign before the angle $\beta_\uparrow$ corresponds to the situation when the slope of the long wave is facing the radar.

Table 5.2 Changes in the length of the Bragg's components of the surface wave field due to changes in the local angle of incidence

Local angle of incidence	$\frac{\lambda_B\left(\theta \pm n\sqrt{\overline{\beta_u^2}}\right)}{\lambda_B(\theta)}$, $n = 1, 2, 3$			
	$\theta = 35°$		$\theta = 45°$	
$\theta \pm \sqrt{\overline{\beta_u^2}}$	0.85	1.2	0.89	1.2
$\theta \pm 2\sqrt{\overline{\beta_u^2}}$	0.75	1.6	0.82	1.4
$\theta \pm 3\sqrt{\overline{\beta_u^2}}$	0.69	2.5	0.77	1.8

The average value $\sqrt{\overline{\xi_u^2}}$ obtained for all experiments, the results of which are shown in Fig. 5.1 is 0.13, which corresponds to $\sqrt{\overline{\beta_u^2}} = 7.3°$. The ranges of the relative variation of the Bragg wavelength at this value $\sqrt{\overline{\beta_u^2}}$ are shown in Table 5.2. The calculations were carried out at an incidence angle of 35° and 45°.

The angular distribution of wave energy in decimeter and longer waves is anisotropic. The variance of sea surface slopes oriented in the wind direction is higher than in the transverse direction. The effect of long waves on the Bragg scattering character of transverse wind direction probing is less than that of transverse wind direction probing. The average value $\sqrt{\overline{\beta_u^2}}$ in the upwind direction is 5.9°. Changes in the local angle of incidence at $\theta = 35°$ correspond to relative changes in the length of the Bragg component $\frac{\lambda_B\left(\theta \pm 3\sqrt{\overline{\beta_c^2}}\right)}{\lambda_B(\theta)}$ in the range from 0.72 to 1.9.

The sea surface elevation spectrum is a steeply falling function of wave numbers, so even a small wavelength change leads to a significant change in the spectral density of the surface waves at the resonant wavelength. Accordingly, as follows from expression (5.6), there is a significant change in the normalized backscatter cross section. When probing along the wind direction at an angle of 45°, changes in the resonance wavelength when the angle changes $\beta_\uparrow$ by a value equal to the mean square of the slope at wind speed of 10 m/s, lead, depending on the sign before $\beta_\uparrow$ in the expression (5.15), to an increase in the normalized cross-section of the backscatter by 89% or a 35% reduction compared with the situation when the resonance waves spread over the horizontal surface.

It should be noted that the effect similar to the change of the resonance wavelength in the presence of long waves takes place at vertical probing (altimeter wind speed measurements). In this case, the boundary between the ranges of surface waves forming mirror reflection and diffuse scattering changes [13], which is one of the factors limiting the limit accuracy of altimeter wind velocity measurements.

5.4.2 Effect of Long Waves on the Bragg Scattering (Gaussian Distribution)

Let us start the analysis of effects created by long waves with consideration of the linear wave field, the slopes of which are subject to the Gauss distribution. For calculations of the normalized backscatter cross-section section, it is necessary to specify changes in the elevation spectrum $S(k)$ at wave number changes. In the range of wave numbers, the lower boundary of which exceeds the wave number of dominant waves, the spectrum is described by approximation $S(k) \sim k^{-n}$. The rate of the spectrum decline, which is determined by the parameter n, depends on the range of wave numbers [8, 13]. In the short-wave region, $n = 3$ can be taken. The boundary separating the gravitational and gravitational-capillary waves is the 7 cm scale [14].

The geometric coefficient $\left|G_{pp}\right|^2$ in Eq. (5.1) depends on the angle of incidence and the physical and chemical properties of seawater. For vertical and horizontal polarization, the geometric coefficient $\left|G_{pp}\right|^2$ is respectively [24]:

$$G_{vv}(\theta, \varepsilon_r) = \cos^4\theta \frac{(\varepsilon_r - 1)\left[\varepsilon_r\left(1 + \sin^2\theta\right) - \sin^2\theta\right]}{\left[\varepsilon_r \cos\theta + \sqrt{\varepsilon_r - \sin^2\theta}\right]^2} \tag{5.16}$$

$$G_{hh}(\theta, \varepsilon_r) = \cos^4\theta \frac{(\varepsilon_r - 1)}{\left[\cos\theta + \sqrt{\varepsilon_r - \sin^2\theta}\right]^2} \tag{5.17}$$

where ε_r is relative dielectric constant of seawater under the ocean-atmosphere interface. The relative dielectric constant of seawater depends on its temperature and salinity, as well as on the frequency of the sounding radio waves [16].

In a situation where probing is performed in the centimeter range, simple approximations obtained in [20] can be used for the geometric coefficient $\left|G_{pp}(\theta)\right|^2$.

$$|G_{vv}|^2 = \frac{\cos^4\theta\left(1 + \sin^2\theta\right)^2}{(\cos\theta + 0.111)^4} \tag{5.18}$$

$$|G_{hh}|^2 = \frac{\cos^4\theta}{(0.111\cos\theta + 1)^4} \tag{5.19}$$

The analysis of the influence of long waves on the Bragg scattering begins with calculations performed under the assumption that the sea surface slopes are subject to Gaussian distribution. After that, let us estimate the effects caused by deviations of the third and fourth statistical moments of the slopes from the values corresponding to the Gaussian distribution.

Move from the probability density of slopes $\tilde{P}(\xi)$ to the probability density of slope angles $P(\beta)$. The procedure for replacing variables is described by the expression

$$P(\beta) = \frac{d\xi}{d\beta} \tilde{P}(tg\beta) \tag{5.20}$$

As a criterion describing the effect of long waves on the Bragg scattering, let us consider the ratio of the backscatter section in the presence of long waves and in the case of Bragg components spreading over a flat surface. Let us introduce the dimensionless parameter

$$\chi_{pp}(\theta) = \frac{\int \sigma_{pp}^{0}(\theta - \beta_{\uparrow}) P(\beta_{\uparrow}) d\beta_{\uparrow}}{\sigma_{pp}^{0}(\theta)} \tag{5.21}$$

The results of calculating the estimates of the parameter $\chi_{pp}^{(G)}$ for the situation where the long waves are longer than 10 cm, are presented in Fig. 5.3. The G index

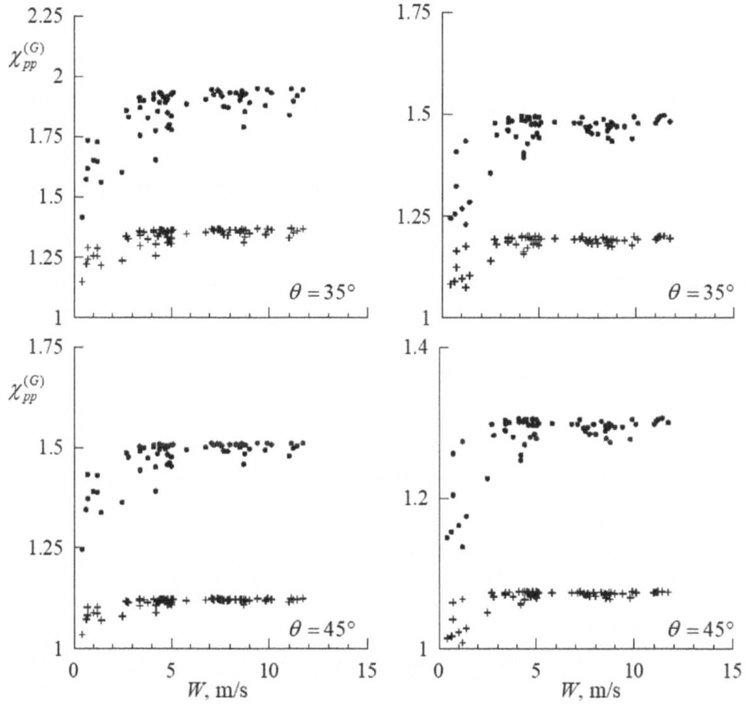

Fig. 5.3 Dependence of the parameter $\chi_{pp}^{(G)}$ on wind speed W: symbol • is horizontal polarization, symbol + is vertical polarization: left column corresponds to probing in the upwind direction, right column corresponds to probing in crosswind direction

has been introduced here, which shows that the calculations were performed for the situation where the distribution of the sea surface slopes is Gaussian.

It follows from the analysis that changes in the local slopes of the sea surface caused by long waves lead to an increase in the backscatter cross-section value. This physical effect takes place both when probing on horizontal and vertical polarization. It is more evident for radars operating on horizontal polarization. The influence of long waves on the backscatter cross section also depends on the angle of incidence of the θ radio waves, with increasing angle of influence of the long waves decreases [27].

5.4.3 *Effects of Quasi-Gaussian Distribution of Sea Surface Slopes*

Numerous experimental studies conducted with different types of equipment have shown that the sea surface slope distribution is not Gaussian [7, 9, 11, 21, 29]. A number of physical mechanisms lead to deviations from Gaussian distribution. The main physical mechanisms are different types of wave interactions of different scales [2, 15].

For modeling the probability density of quasi-Gaussian distributions, approximations based on the Gram-Charlier series are used [9]. Coefficients of the Gram-Charlier series are calculated on the basis of empirical estimations of statistical moment slope. The determination of senior statistical moments in marine conditions is associated with known technical difficulties [19, 21, 28]. Usually, only statistical moments up to and including the fourth order of magnitude are determined in experiments, so only the first five members of the Gram-Charlier series are taken into account when modeling the statistical distribution of slopes.

After the introduction of normalization $\tilde{\xi} = \xi/\sqrt{\overline{\xi^2}}$, the approximation of the probability density function of the slope components, for which skewness and kurtosis are known, can be presented in a simple form

$$P_{G-C}\left(\tilde{\xi}\right) = \frac{1}{\sqrt{2\pi}} \exp\left(-\frac{\tilde{\xi}^2}{2}\right)\left(1 + \frac{A}{6} H_3\left(\tilde{\xi}\right) + \frac{E}{24} H_4\left(\tilde{\xi}\right)\right) \tag{5.22}$$

where H_3 and H_4 Hermit polynomials third and fourth order, $H_3\left(\tilde{\xi}\right) = \tilde{\xi}^3 - 3\tilde{\xi}$ and $H_4\left(\tilde{\xi}\right) = \tilde{\xi}^4 - 6\tilde{\xi}^2 + 3$.

The values of the module of skewness of the upwind and crosswind components of the slopes of the sea surface created by waves longer than 10 cm, as shown in Fig. 5.1, do not exceed level 0.3. The values of the kurtosis mainly lie between −0.4 and 0.5. There is no correlation between skewness and kurtosis. The type of the probability density $P_{G-C}\left(\tilde{\xi}\right)$ functions constructed for skewness and kurtosis values

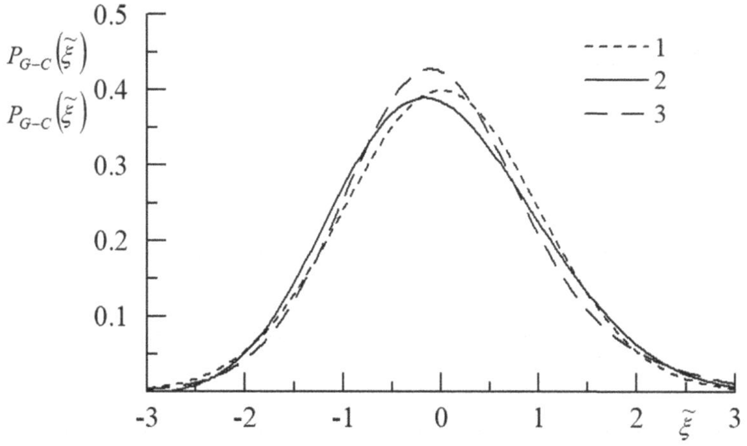

Fig. 5.4 Probability density functions: curve 1 is Gaussian distribution; curves 2, 3 is distributions built on the basis of truncated Gram-Charlier series, $A = 0.3$, and $E = 0.5A = 0.3E = -0.3$, respectively

corresponding to significant deviations from the Gaussian distribution, but lying in the specified ranges, is presented in Fig. 5.4.

Deviations in the slope distribution from the Gaussian distribution should lead to a change in the parameter $\chi_{pp}^{(G-C)}$. Here, the upper index *G-H* means that calculations of the influence of long waves according to the formula (5.21) are performed for the distribution of the sea surface elevations in a form (5.22). The dependence of the parameter $\chi_{pp}^{(G-C)}$ on skewness changes is not unambiguous, it depends on the direction in which the probing is performed relative to the wind velocity vector. If probing is carried out upwind, the increase in skewness leads to growth $\chi_{pp}^{(G-C)}$, if in the downwind direction $\chi_{pp}^{(G-C)}$ decreases.

Let us consider the parameter ε_{pp} that allows us to estimate the extent to which the deviations of slope distributions from the Gaussian distribution affect the value of the backscatter radar cross section

$$\varepsilon_{pp} = \chi_{pp}^{(G-C)} / \chi_{pp}^{(G)} \tag{5.23}$$

Changes in the parameter ε_{pp} when changing the skewness and the slope kurtosis are shown in Fig. 5.5. Here the negative values of skewness correspond to probing in the wind direction. Calculations were made for the probing angle of $\theta = 35°$ and $\sqrt{\overline{\beta_{\uparrow}^2}} = 7.3°$. With different polarization types, deviations from Gaussian distribtion have different effect on the backscatter radar cross section σ_{pp}^{L}. Their influence is stronger if the polarization is horizontal. Changes in the distribution kurtosis of the sea surface slopes within the scattering of individual values determined in the experiment (see Fig. 5.1) have little effect on the value of the parameter ε_{pp}.

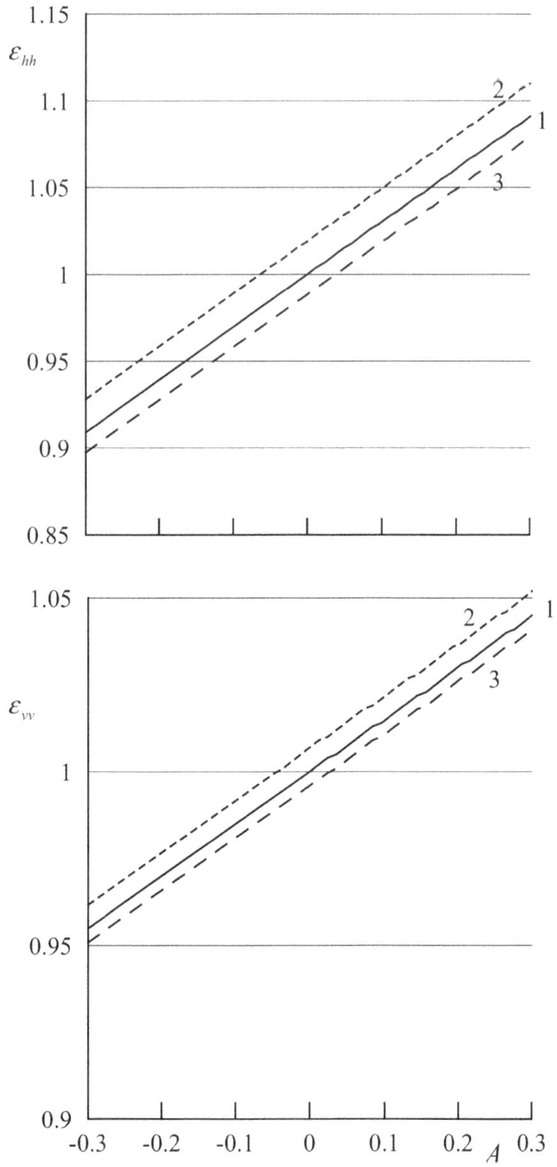

Fig. 5.5 Dependence of the parameter ε_{pp} on the skewness of A and the kurtosis of E components of the sea surface slopes: curve 1 is $E = 0$; curve 2 is $E = 0.5$; curve 3—$E = 0.3$

Studies of sea surface slopes show strong variability of their statistical moments even at fixed wind speeds [29]. The skewness and kurtosis of the slope are weakly correlated with each other. It is interesting to consider how the values of the parameter ε_{pp} change in individual situations. The results of calculations of the parameter ε_{pp} at different wind speeds are presented in Fig. 5.6.

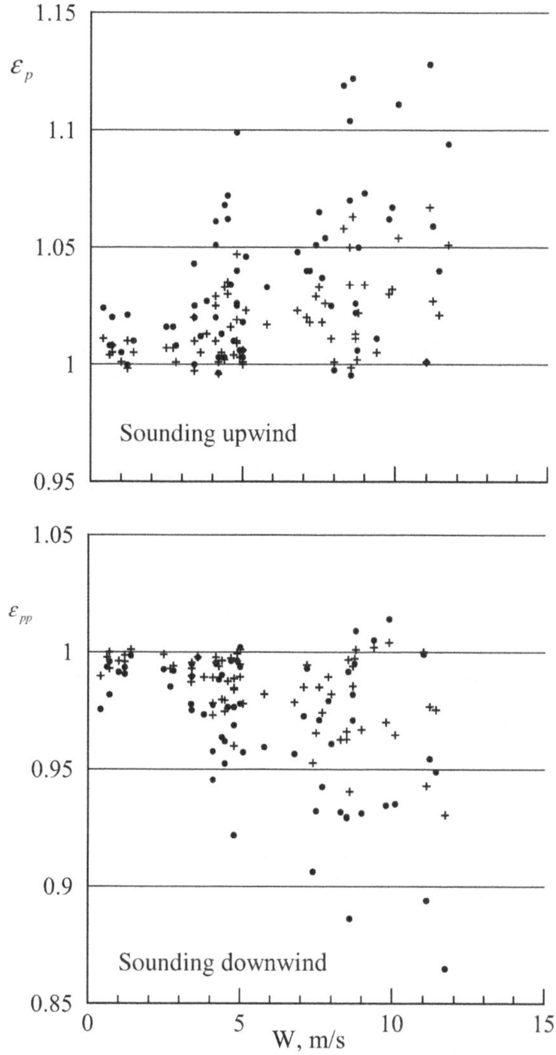

Fig. 5.6 Wind speed W dependencies of the parameter ε_{pp}, based on empirical estimates of skewness and kurtosis of the upwind component of slopes: • correspond to horizontal polarization, + correspond to vertical polarization

Calculations were carried out in two variants, corresponding to probing in the wind direction and in the direction of the wind. In case of horizontal polarization, skewness of slope distributions can lead to a deviation of the backscatter section up to 15% of the values obtained, provided that $A = 0$. Calculations were made at $\theta = 35°$. For vertical polarization, the deviation from the backscattering cross section estimates obtained for Gaussian distributions is approximately half as small [27].

5.5 Sounding of the Sea Surface in the Millimetre Range of Radio Wave

Let us consider a situation when, in relation to the resonance components, the long waves are those in the centimeter range and longer, which corresponds to radio probing in the millimeter range. In this case, we can no longer use Eqs. (5.18) and (5.19) to calculate the geometric factor $\left|G_{pp}(\theta)\right|^2$, we must use Eqs. (5.16) and (5.17).

Let us consider how the geometric factor changes when probing at radio wavelengths equal to Λ 0.2 and 0.8 cm. The dependencies of the geometric factor on the angle of incidence are shown in Fig. 5.7. Calculations were made for salinity values characteristic of the Black Sea, S = 17‰, and water temperature was assumed to be 15 °C. We can see that the curves describing the geometric coefficients $\left|G_{pp}(\theta)\right|^2$, when the wavelength changes, differ not only in level but also in slope. The slope change of the curve, when the radio wavelength changes, is higher for horizontal polarization than for vertical polarization.

The greater the slope of the curve $\left|G_{pp}(\theta)\right|^2$, the greater is the effect of changes in local slopes created by long waves on the level of the backward scattered signal. To estimate the changes in the slope of the curve with changes in the incidence angle at different wavelengths, let us introduce a parameter that characterizes the sensitivity of the geometric coefficient to changes in the incidence angle.

$$R_{pp}(\theta) = \Delta G_{pp}(\theta) \Big/ \overline{\left|G_{pp}(\theta)\right|^2} \tag{5.24}$$

where $\Delta G_{pp}(\theta)$ is the difference of geometric coefficients values when the angle of incidence changes by 1°; $\overline{\left|G_{pp}(\theta)\right|^2}$ is average values of geometric coefficients in the range of incidence angle change by 1°.

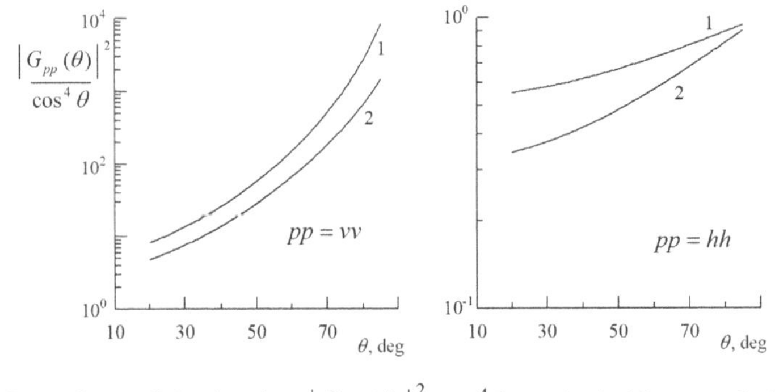

Fig. 5.7 Dependence of the function $\left|G_{pp}(\theta)\right|^2 \cos^{-4}\theta$ on the incidence angle θ. Curve 1 corresponds to Λ = 0.8 cm, curve 2 corresponds to Λ = 0.2 cm

The sensitivity of the geometric factor to changes in the angle of incidence, which is characterized by the parameter $R_{pp}(\theta)$, when probing on the vertical polarization is higher than on the horizontal. However, with vertical polarization, when the wavelength changes, it changes much weaker than with horizontal polarization. Noticeable change in sensitivity $R_{VV}(\theta)$ when changing the radio wavelength takes place only at angles of incidence exceeding 65°. On the horizontal polarization, the sensitivity decreases by 80% when the wavelength changes from 0.2 to 0.8 cm. Moreover, these changes practically do not depend on the angle of incidence.

5.6 Conclusion

The influence of sea surface waves longer than Bragg components on the backscatter of centimeter and millimeter radio waves has been analyzed. Two physical mechanisms leading to a change in the backscattering cross-section in the presence of long waves are analyzed, the change in the local incidence angle and the change in the Bragg wavelength. Situations where long waves are the surface waves longer than 10 cm and longer than 2 cm are considered.

Comparing the results presented above with those published in recent years on sea surface radio wave scattering, it should be noted that their principal difference is that they are obtained from calculations based on direct in situ measurements of sea surface slopes. This made it possible to estimate the influence of real deviations of sea surface slope distribution from Gauss distribution.

Changes in the local slopes of the sea surface result in an increased backscatter section. It is shown that the sensitivity of the geometric factor to changes caused by long waves of the local incidence angle for probing on vertical polarization is higher than on horizontal polarization.

The spread at a given wind velocity of the statistical characteristics of the sea surface slopes created by surface waves longer than the Bragg components is one of the factors limiting the ultimate accuracy of the scatterometer determination of wind velocity.

References

1. Apel JR (1994) An improved model of the ocean surface wave vector spectrum and its effects on radar backscatter. J Geophys Res 99(C8):16269–16291
2. Babanin AV, Polnikov VG (1994) On non-Gaussian wind waves. Phys Oceanogr 3:79–82
3. Bakhanov VV, Demakova AA, Korinenko AE, Ryabkova MS, Titov VI (2018) Estimation of the wind wave spectra with centimeters-to-meter lengths by the sea surface images. Phys Oceanogr 25(3):177–190
4. Bass FG, Fuks IM (1979) Wave scattering by statistically rough surface. Pergamon, 540 p
5. Bass FG, Braude SYa, Kalmykov AI, Men' AV, Ostrovskii IY, Pustovoitenko VV, Rozenberg AD, Fuks IM (1975) Radar methods for the study of ocean waves (Radiooceanography). Sov Phys Usp 18:641–642. https://doi.org/10.1070/pu1975v018n08abeh004920

6. Berginc G (2002) Small-slope approximation method: a further study of vector wave scattering from two-dimensional surfaces and comparison with experimental data. Prog Electromagn Res PIER 37:251–287
7. Bréon FM, Henriot N (2006) Spaceborne observations of ocean glint reflectance and modeling of wave slope distributions. J Geophys Res 111(6):C06005
8. Cheng Y, Liu Y, Xu Q (2006) A new wind-wave spectrum model for deep water. Indian J Mar Sci 35(3):181–194
9. Cox C, Munk W (1954) Measurements of the roughness of the sea surface from photographs of the sun glitter. J Opt Soc Am 44(11):838–850
10. Hughes BA, Grant HL, Chappell RWA (1977) A fast response surface-wave slope meter and measured wind-wave components. Deep-Sea Res 24(12):1211–1223
11. Khristophorov GN, Zapevalov AS, Babiy MV (1992) Statistics of sea-surface slope for different wind speeds. Okeanologiya 32(3):452–459
12. Konstantinov OG, Novotryasov VV (2013) Surface manifestations of internal waves observed using a land-based video system. Izvestiya Atmos Ocean Phys 49(3):334–338
13. Kudryavtsev VN, Hauser D, Caudal G, Chapron BA (2002) A semi-empirical model of the normalized radar cross-section of the sea surface. 1. Background model. J Geophys Res 107:XXXX. https://doi.org/10.1029/2001jc001003
14. Lighthill J (2001) Waves in fluids. Reissued in the Cambridge mathematical library series, 524 p
15. Longuet-Higgins MS (1963) The effect of non-linearities on statistical distribution in the theory of sea waves. J Fluid Mech 17(3):459–480
16. Meissner T, Wentz FJ (2004) The complex dielectric constant of pure and sea water from micro-wave satellite observations. IEEE Trans Geosci Remote Sens 42(9):1836–1849
17. Moore RK, Pierson WJ (1966) Measuring sea state and estimating surface winds from a polar orbiting satellite. In: Proceedings of international symposium on electromagnetic sensing of the earth from satellites, Miami Beach, FL, 22–24 Nov 1966, pp R1–R28
18. Nieto Borge JC, Rodriguez GN, Hessner K, Gonzalez PI (2004) Inversion of marine radar images for surface wave analysis. J Atmos Ocean Technol 21:1291–1300
19. Palm CS, Anderson RS, Reese AM (1977) Laser probe for measuring 2-D wave slope spectra of ocean capillary waves. J Appl Opt 46(4):1074–1081
20. Plant WJ (1986) A two-scale model of short wind generated waves and scatterometry. J Geophys Res 91(C9):10735–10749
21. Pustovoytenko VV, Lebedev NE (2015) Comparison of sea surface slope statistical moments obtained by means of optical scanners and laser inclinometers. Sovremennye Problemy Distantsionnogo Zondirovaniya Zemli iz Kosmosa 12(1):102–109
22. Ratner YuB, Fomin VV, Ivanchik AM, Ivanchik MV (2017) System of the wind wave operational forecast in the Black Sea marine forecasting center. Phys Oceanogr 5:51–59
23. Thompson D, Elfouhaily T, Chapron B (1998) Polarization ratio for microwave backscattering from the ocean surface at low to moderate incidence angles. In: Geoscience and remote sensing symposium proceedings, 1998. IGARSS '98. https://doi.org/10.1109/igarss.1998.692411
24. Valenzuela G (1978) Theories for the interaction of electromagnetic and ocean waves—a review. Bound Layer Meteorol 13(1–4):61–85
25. Walsh EJ, Banner ML, Wright CW, Vandemark DC, Chapron B, Jensen J, Lee S (2008) The southern ocean waves experiment. Part III: Sea surface slope statistics and near-nadir remote sensing. J Phys Oceanogr 38:670–685
26. West JC, O'Leary BS, Klinke J (1998) Numerical calculation of electromagnetic scattering from measured wind-roughened water surfaces. Int J Remote Sens 19(7):1377–1393
27. Zapevalov AS (2009) Bragg scattering of centimeter electromagnetic radiation from the sea surface: the effect of waves longer than Bragg components. Izvestiya Atmos Ocean Phys 45(2):253–261
28. Zapevalov AS (2018) Determination of the statistical moments of sea-surface slopes by optical scanners. Atmos Ocean Opt 31(1):91–95. https://doi.org/10.1134/S1024856018010141

29. Zapevalov AS (2002) Statistical characteristics of the moduli of slopes of the sea surface. Phys Oceanogr 12(1):24–31
30. Zapevalov AS (2008) Statistical models of the sea surface in problems of acoustic and electromagnetic radiation scattering. Manuscript to claim the academic degree of doctor of physico-mathematical sciences on the speciality 04.00.22—geophysics. Marine Hydrophysical Institute of the National Academy of Sciences of Ukraine, Sebastopol
31. Zapevalov AS, Lebedev NE (2014) Simulation of statistical characteristics of sea surface during remote optical sensing. Atmos Ocean Opt 27(6):487–492
32. Zavorotny VU, Voronovich AG (1998) Two scale model and ocean radar Doppler spectra at moderate and low grazing angles. IEEE Trans Antennas Propag 46(1):84–92

Chapter 6
Impact of Physicochemical Characteristics of the Sea Water on Bregg Scattering of Radio Waves

6.1 Introduction

There is now a steady increase in the amount of information on atmospheric and oceanic processes obtained from satellite-based remote sensing instruments. As the volume of information grows, there are qualitative changes, the number of remotely sensed parameters increases. In radar sensing, almost all information is "read out" from the sea surface. Therefore, for correct interpretation of sensing results it is necessary to have a detailed view of the role of individual physical mechanisms that form the signal recorded on a spacecraft, including changes in the physical and chemical characteristics of the sea surface layer [8, 17, 28].

The main source of information on the physical and chemical characteristics of the surface layer of the World Ocean is satellite microwave radiometers [9]. The output signals of microwave radiometers are proportional to the emissivity of the sea surface, which is small, which imposes limitations on their accuracy and resolution [1].

The development of microwave radiometry takes place in several directions: the frequency range of measured thermal radio radiation is extended, high-sensitive receiving equipment is developed, spatial resolution is increased by changing the receiving antenna [23]. Another recently proposed approach is the use of active radar methods [32]. This approach is based on the dependence of the level of the reflected radio signal from the sea surface on the physical and chemical characteristics of the water under the ocean-atmosphere boundary.

When probing the sea surface in the microwave range in the region of incidence angles, where the resonance scattering mechanism dominates, the signal recorded is proportional to the geometric coefficient dependent on the incidence angle of the radio waves and the complex dielectric constant of the medium under the atmosphere-ocean boundary [2, 29]. Theoretical and experimental studies have shown that there

A. Zapevalov et al., *Simulation of the Sea Surface for Remote Sensing*, Springer Oceanography,
https://doi.org/10.1007/978-3-030-58752-9_6

is a dependence of the relative dielectric permeability of sea water on its temperature and salinity [10, 26]. This effect has a noticeable effect on the results of monitoring the boundary layers of the atmosphere and ocean by active remote sensing of the microwave range [28].

Here considers the influence of physical and chemical characteristics of the marine environment on changes in the level of radio signal scattered back by surface waves, analyzes the possibility of measuring the temperature and salinity of sea water by means of active radiosonding.

6.2 Dependence of the Complex Dielectric Constant of Water on Its Temperature and Salinity

The complex dielectric constant of seawater is a function of three parameters: temperature, salinity and frequency. Active experimental studies of this characteristic began in the middle of the last century [13]. In recent decades, this research has again attracted close attention, which is associated with the development of radiometric means of controlling the temperature and salinity of the oceans [25].

The accuracy of recovery of marine environmental parameters depends on the quality of the model of complex dielectric constant used. In the present work we will use the model describing the relationship between complex dielectric constant of sea water and temperature, salinity, and frequency, which is proposed in work [19].

The model is based on Debai's two-frequency approximation.

$$\varepsilon(T,S)=\frac{\varepsilon_S(T,S)-\varepsilon_1(T,S)}{1+iw/w_1(T,S)}+\frac{\varepsilon_1(T,S)-\varepsilon_\infty(T,S)}{1+iw/w_2(T,S)}+\varepsilon_\infty(T,S)-i\frac{\sigma(T,S)}{2\pi\varepsilon_0 w} \tag{6.1}$$

where T is temperature; S is salinity; w is the frequency of the radio wave;w_1 и w_2 are the first and second frequencies of Debai; ε_S is the static dielectric constant; $i=\sqrt{-1}$; $\sigma(T,S)$ is the conductivity of sea water (Sim/m), ε_0 is the vacuum electric permittivity. The characteristics defining $\varepsilon(T,S)$ are described by the equations

$$1/(2\pi\varepsilon_0)=17.97510\ (\text{GHz m/S}) \tag{6.2}$$

$$\varepsilon_S(T,S=0)=\frac{3.70886\times10^4-8.2168\cdot10^1T}{4.2154\times10^2+T} \tag{6.3}$$

$$w_1(T,S=0)=\frac{45+T}{a_3+a_4T+a_5T^2} \tag{6.4}$$

$$\varepsilon_\infty(T,S=0)=a_6+a_7T \tag{6.5}$$

$$w_2(T, S = 0) = \frac{45 + T}{a_8 + a_9 T + a_{10} T^2} \tag{6.6}$$

$$\varepsilon_S(T, S) = \varepsilon_S(T, S = 0) \exp\left[b_0 S + b_1 S^2 + b_2 T S\right] \tag{6.7}$$

$$w_1(T, S) = v_1(T, S = 0)\left[1 + S\left(b_3 + b_4 T + b_5 T^2\right)\right] \tag{6.8}$$

$$\varepsilon_1(T, S) = \varepsilon_1(T, S = 0) \exp\left[b_6 S + b_7 S^2 + b_8 T S\right] \tag{6.9}$$

$$w_2(T, S) = v_2(T, S = 0)\ [1 + S(b_9 + b_{10} T)] \tag{6.10}$$

$$\varepsilon_\infty(T, S) = \varepsilon_\infty(T, S = 0)\ [1 + S(b_{11} + b_{12} T)] \tag{6.11}$$

The numerical values of the coefficients a_i and b_i included in Eqs. (6.3)–(6.10) are shown in Table 6.1.

In [19] for calculations of electroconductivity of sea water regression dependence was used [27], obtained on the basis of laboratory experiments

$$\sigma(T, S) = \sigma(T, S = 35)\ R_{15}(S) \frac{R_T(S)}{R_{15}(S)} \tag{6.12}$$

where

$$\sigma(T, S = 35) = 2.903602 + 8.607 \times 10^{-2} T + 4.738817 \times 10^{-4} T^2$$

Table 6.1 Coefficients a_i and b_i

i	a_i	b_i
0	5.7230	-3.56417×10^{-3}
1	2.2379×10^{-2}	4.74868×10^{-6}
2	$-7.1237\ 10^{-4}$	1.15574×10^{-5}
3	5.0478	2.39357×10^{-3}
4	-7.0315×10^{-2}	-3.13530×10^{-5}
5	6.0059×10^{-4}	2.52477×10^{-7}
6	3.6143	-6.28908×10^{-3}
7	2.8841×10^{-2}	1.76032×10^{-4}
8	1.3652×10^{-1}	-9.22144×10^{-5}
9	1.4825×10^{-3}	-1.99723×10^{-2}
10	2.4166×10^{-4}	1.81176×10^{-4}
11		-2.04265×10^{-3}
12		1.57883×10^{-4}

$$- 2.991 \times 10^{-6}T^3 + 4.3047 \times 10^{-9}T^4 \tag{6.13}$$

$$R_{15}(S) = S\frac{37.5109 + 5.45216 \cdot S + 1.4409 \times 10^{-2}S^2}{1004.75 + 182.283\,S + S^2} \tag{6.14}$$

$$\frac{R_T(S)}{R_{15}(S)} = 1 + \frac{\alpha_0(T - 15)}{\alpha_1 + T} \tag{6.15}$$

$$\alpha_0 = \frac{6.9431 + 3.2841\,S - 9.9486 \times 10^{-2}S^2}{84.850 + 69.024\,S + S^2} \tag{6.16}$$

$$\alpha_1 = 49.843 - 0.2276\,S + 0.198 \times 10^{-2}S^2 \tag{6.17}$$

Temperature T is degree Celsius (°C); salinity S is expressed in units of parts per thousand (ppt or ‰), frequencies w, w_1 and w_2 are expressed in gigahertz (GHz), conductivity of water is $\sigma(T, S)$ expressed in siemens per meter (S/m).

The dependencies of the complex dielectric constant ε on frequency w at three salinity values and two temperature values are shown in the Fig. 6.1. It can be seen that the imaginary part is ε more sensitive to changes in temperature and salinity

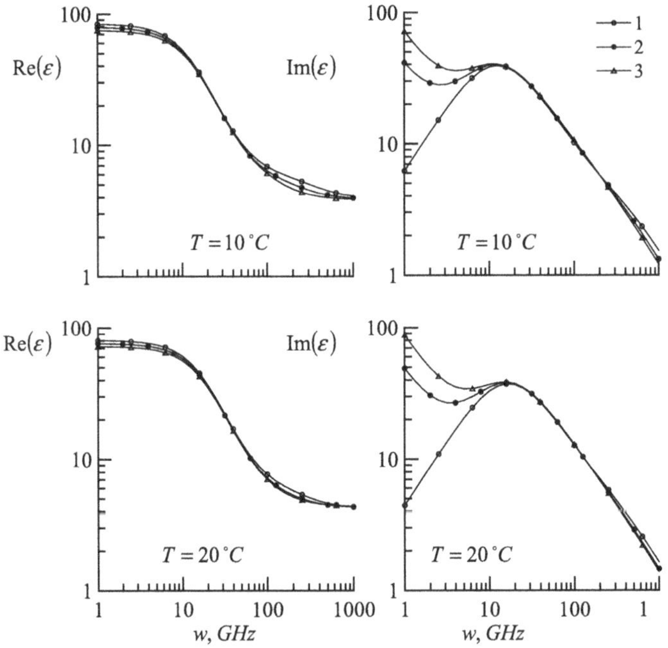

Fig. 6.1 Frequency dependencies w of real Re(ε) and imaginary Im(ε) components of dielectric permeability, constructed for two temperature values, $T = 10\,°C$ and $T = 20\,°C$. Curves 1–3 correspond to salinity 0‰, 17‰, 34‰

than the real part ε. At low frequencies, this dependence is expressed more strongly than at high frequencies. Also, from Fig. 6.1 shows that the nature of the dependence $\varepsilon = \varepsilon(w, T, S)$ is strongly frequency dependent at all T and S values.

For further analysis of the Bregg scattering of radio waves on the sea surface we will move from the frequency space w to the wavelength space Λ. These parameters are related by the ratio $\Lambda = c/w$ where c is the speed of light. In Fig. 6.2 presents the dependencies on the temperature of real $\mathrm{Re}(\varepsilon)$ and imaginary $\mathrm{Im}(\varepsilon)$ parts of the relative dielectric permeability of sea water at the three frequencies 37.5, 10 and 3 GHz, corresponding to radio wavelengths $\Lambda = 0.8$, 3 and 10 cm. Calculations were made at a salinity of 17‰, which is close to the Black Sea salinity.

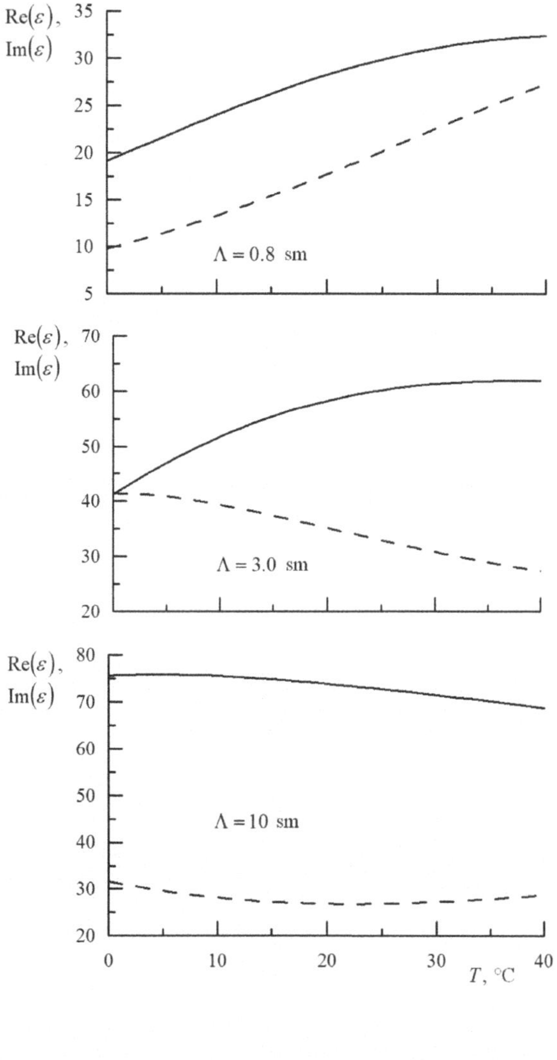

Fig. 6.2 Dependence on temperature T of the actual $\mathrm{Re}(\varepsilon)$ (solid line) and imaginary $\mathrm{Im}(\varepsilon)$ (dashed line) parts of the complex dielectric constant

Similar calculations of dependence Re(ε) and Im(ε) salinity at fixed temperature showed that at shorter wavelengths the sensitivity to changes in salinity is higher than at longer wavelengths. The dependencies Re(ε) and Im(ε) on fixed temperature salinity ($T = 15$ °C) are also shown in Fig. 6.3.

The range of wavelength variation from 0.8 to 10 cm is chosen to cover the range in which active ocean remote sensing radars normally operate.

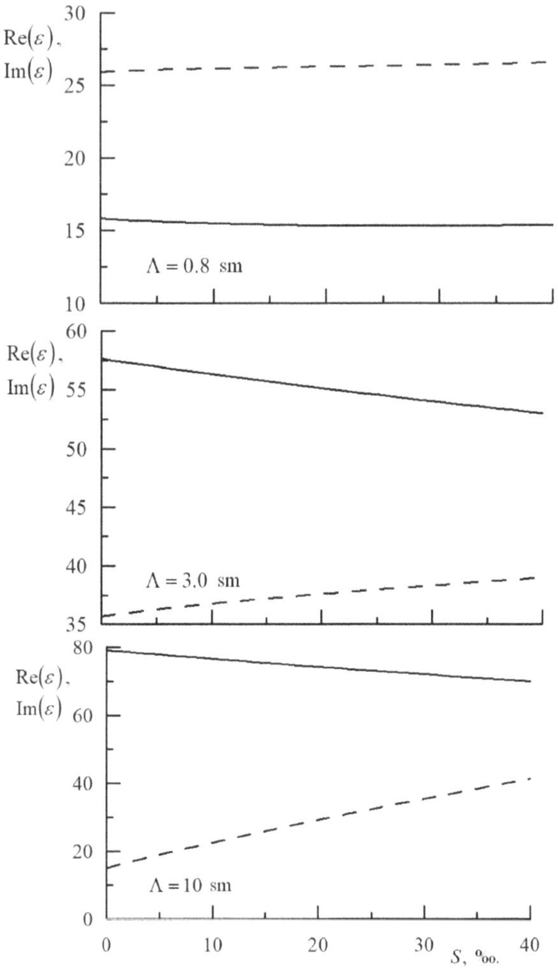

Fig. 6.3 Dependence on the salinity S of the actual Re(ε) (solid line) and imaginary Im(ε) (dashed line) parts of the complex dielectric constant

6.3 Bragg Scattering Radio Wave

If the sounding of the sea surface is carried out angles of incidence θ from 25 ° to 75°–80° backscattering signal determines Bregg (resonance) scattering mechanism [2]. The angle of incidence is counted from the vertical. Breggs backscattering radio waves create surface waves running along the direction of sensing in the forward or backward direction, the wave number. The resonance condition relates the surface wavelength λ_B to the wavelength of the radio waves Λ

$$\lambda_B = \Lambda \big/ (2 \sin \theta) \tag{6.18}$$

In a zero approximation, when the Bregg components of the surface waves spread over a flat surface, the backscatter section can be represented in form [15]

$$\sigma_{pp}^0 = 8\,\pi\, K^4 \cos^4 \theta \left| G_{pp}\,(\theta, \varepsilon) \right|^2 \left[\Xi\left(\vec{k}_B\right) + \Xi\left(-\vec{k}_B\right) \right] \tag{6.19}$$

where pp is type of polarization, the first index corresponds to the polarization of the emitted signal, the second is the received; $\left|G_{pp}\,(\theta)\right|^2$ is the Bragg scattering geometric coefficient, depending on the type of polarization of the emitted and received radar signal and the complex dielectric constant of sea water; $\Xi\,(\vec{k}_B)$ is the spectrum of sea surface waves, on the scale of the Bregg component.

Let's consider a situation in which the radiation and reception of the radio signal are on the same polarization. For vertical and horizontal polarization, the geometric coefficient has the form accordingly ($G_{pp}(\theta, \varepsilon)$ [29]

$$G_{vv}(\theta, \varepsilon) = \cos^2 \theta \frac{(\varepsilon - 1)\left[\varepsilon\left(1 + \sin^2\theta\right) - \sin^2\theta\right]}{\left[\varepsilon\,\cos\theta + \sqrt{\varepsilon - \sin^2\theta}\right]^2} \tag{6.20}$$

$$G_{hh}(\theta, \varepsilon) = \cos^2 \theta \frac{(\varepsilon - 1)}{\left[\cos\theta + \sqrt{\varepsilon - \sin^2\theta}\right]^2} \tag{6.21}$$

where the indexes v and h indicate vertical and horizontal polarization respectively.

Thus, it follows from Eqs. (6.19) to (6.21) that the backscattered radio signal carries information about the roughness level as well as the complex dielectric constant, which in turn depends on temperature and salinity.

6.4 Dependence of Geometric Coefficient on Water Temperature and Salinity

The general view of the incidence angle dependence geometric coefficient, calculated for the three wavelengths on the vertical and horizontal polarization, is shown in the Fig. 6.4. On the vertical and horizontal polarization, the $G_{pp}(\theta, \varepsilon)$ calculated for the wavelengths $\Lambda_1 = 0.8$ cm, $\Lambda_2 = 3$ cm and $\Lambda_3 = 10$ cm are close to each other. It can also be seen that the range of variation $|G_{vv}(\theta)|^2$ when the incidence angle changes is much wider than the range of variation $|G_{hh}(\theta)|^2$.

Note that although the geometric coefficients $G_{pp}(\theta, \varepsilon)$ are defined for incidence angle θ from 0° to 90°, changes in a narrower range are of practical interest. This range is 25° to $< \theta <$ 80°. At small incidence angle the radio signal reflected from the sea surface forms a quasi-mirror reflection mechanism [3, 29]. When probing at incidence angle of $\theta > 80°$, it is necessary to take into account a number of physical effects [12], whose contribution to the radio signal can be comparable with that of the Bregg scattering. In particular, at large incidence angles, the effect of shading of the sea surface by ridges of long dominant waves must be taken into account.

To quantify the difference between the geometric coefficients at the selected wavelengths, let us introduce the coefficients

$$N1_{pp}(\theta) = \left|G_{pp}(\Lambda_1, \theta)^2\right| / \left|G_{pp}(\Lambda_2, \theta)^2\right|, \tag{6.22}$$

$$N2_{pp}(\theta) = \left|G_{pp}(\Lambda_3, \theta)^2\right| / \left|G_{pp}(\Lambda_2, \theta)^2\right|. \tag{6.23}$$

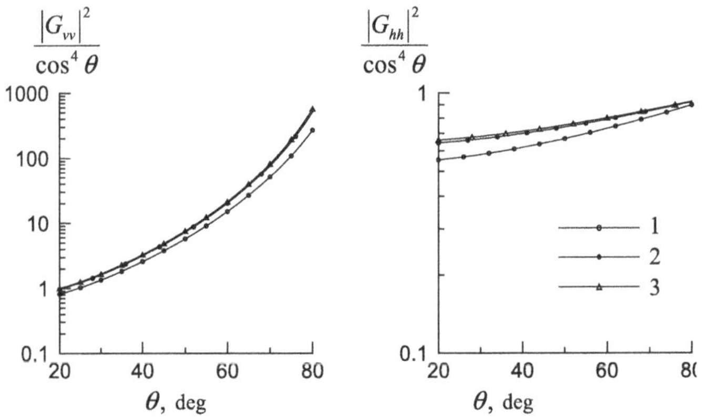

Fig. 6.4 Dependent functions and $|G_{vv}|^2 \cos^{-4}\theta |G_{hh}|^2 \cos^{-4}\theta$ incidence angle θ. Curves 1–3 correspond to radio wavelengths $\Lambda_1 = 0.8$ cm, $\Lambda_2 = 3$ cm and $\Lambda_3 = 10$ cm

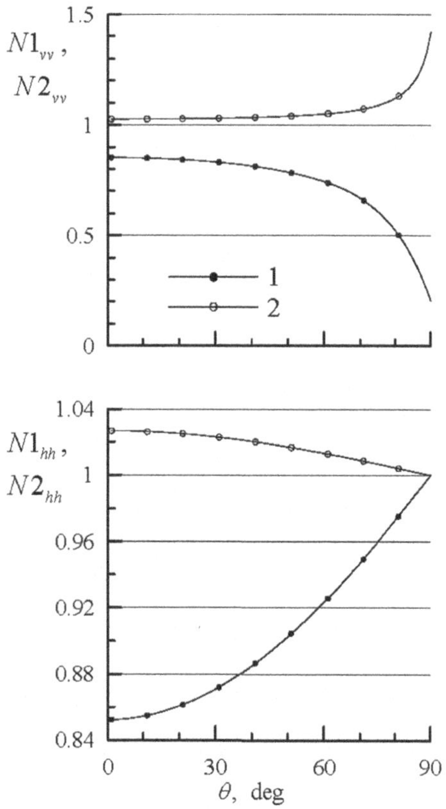

Fig. 6.5 Dependence of coefficients $N1_{pp}(\theta)$ (curve 1) and $N2_{pp}(\theta)$ (curve 2) on the incidence angle θ

where $\Lambda_1 = 0.8$ sm, $\Lambda_2 = 3$ sm, $\Lambda_3 = 10$ sm. In the area $\theta < 60°$ of the divergence of geometric coefficients is $|G_{vv}|^2$ less than 25%, the divergence of geometric coefficients in the $|G_{hh}|^2$ whole range of variation of the angle of incidence is less than 10% (Fig. 6.5).

Let's analyze the dependence of coefficients $\left|G_{pp}(\theta)^2\right|$ at fixed values of incidence angle on salinity and temperature. We will consider three incidence angles $\theta = 30°$, 55° and 80°. We will introduce relative changes of geometric coefficients describing their dependence on salinity and temperature.

$$\tilde{G}_{ppS} = \frac{\left|G_{pp}(\Lambda, \theta, S, T)\right|^2}{\left|G_{pp}(\Lambda, \theta, S_0 = const, T)\right|^2} \tag{6.24}$$

$$\tilde{G}_{ppT} = \frac{\left|G_{pp}(\Lambda, \theta, S, T)\right|^2}{\left|G_{pp}(\Lambda, \theta, S, T_0 = const)\right|^2} \tag{6.25}$$

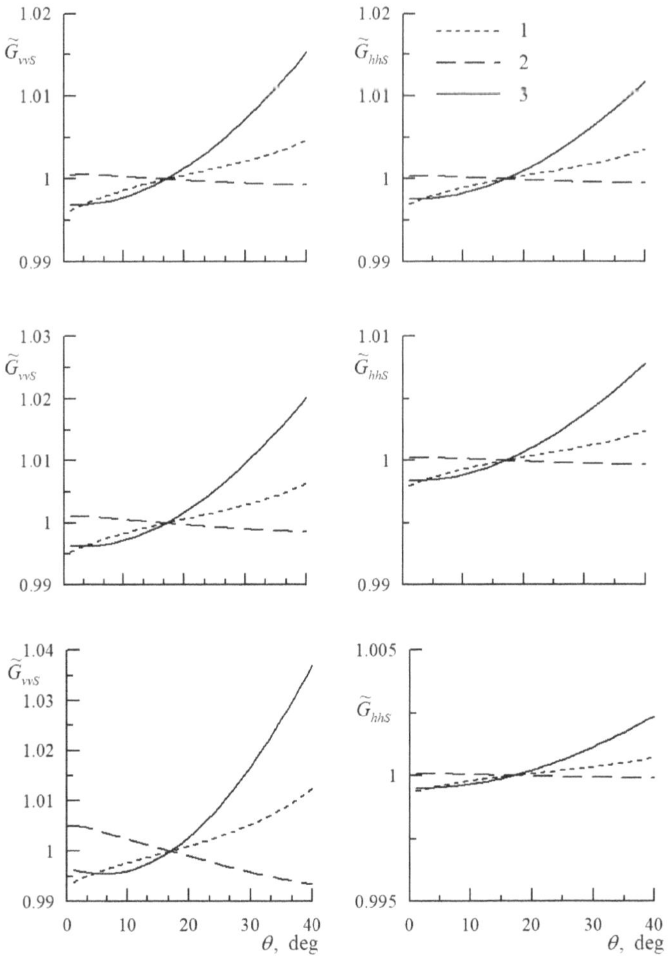

Fig. 6.6 Dependence of parameters $\tilde{G}_{vvS}$ and $\tilde{G}_{hhS}$ on angle of incidence

The results of calculation of parameters $\tilde{G}_{ppS}$ and $\tilde{G}_{ppT}$ are shown in Figs. 6.6 and 6.7 respectively.

Estimations are received at values of salinity $S_0 = const = 17‰$, for temperature $T = 15$ °C. It is seen (see Fig. 6.6) that changes in salinity in a wide range have little effect on the value of the parameter $\tilde{G}_{ppS}$. Salinity changes in the range from 0‰ (fresh water) to 35‰ (ocean water). Changes in the values of the parameter $\tilde{G}_{ppS}$ on both polarizations do not exceed 4%, relative to the value obtained by the reference salinity of 17‰.

Changes in the values of the parameter $\tilde{G}_{ppT}$ occurs within a wider range than the parameter $\tilde{G}_{ppS}$, which follows from the comparison of Figs. 6.6 and 6.7. Here the temperature value $T_0 = 15$ °C is taken as a reference. Comparing the graphs

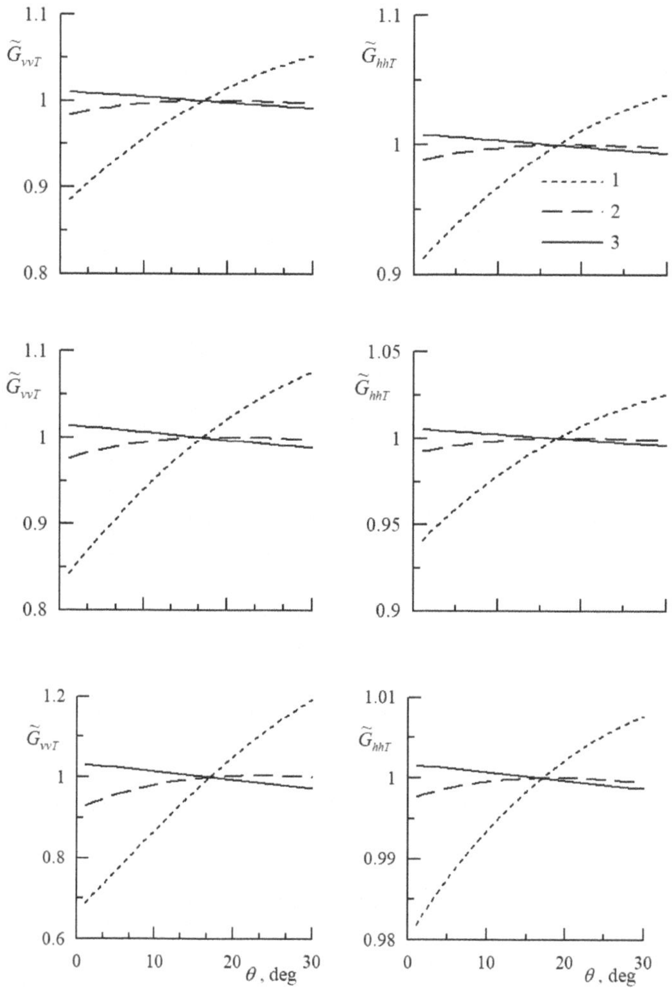

Fig. 6.7 Dependence of parameters $\tilde{G}_{vvT}$ and $\tilde{G}_{hhT}$ on angle of incidence

$\tilde{G}_{ppT}$, plotted for different angles of incidence, we can conclude that with increasing angle of incidence the θ dependence of the parameter $\tilde{G}_{ppT}$ on temperature increases. Changes in the values of the parameter $\tilde{G}_{vvT}$ for the 80° angle of incidence can reach 25%.

It follows from Eq. (6.18) that the variability of the normalized backscatter section σ_{pp}^0 depends on the variability of the surface wave spectrum on the Bregg component scale. Let us determine the relative changes in the wave spectra on those scales that determine the microwave radio wave scattering. For this purpose, we use the unidirectional spectrum model $\tilde{\Psi}(k)$ proposed in the paper [18].

$$\tilde{\Psi}(k) = \int_0^{2\pi} \tilde{\Psi}(k, \alpha) d\alpha \tag{6.26}$$

where $k = \left|\vec{k}\right|$ is wave number; α is azimuth angle.

The spectra of sea surface elevations fall rapidly with wave number k growth [5, 15]. Therefore, curvature spectra are widely used in their analysis. The curvature of the sea surface is the second spatial derivative of its elevations, respectively, the spectrum of curvature is related to the spectrum of elevations by the ratio $Cr(k) = k^4 \tilde{\Psi}(k)$.

The curvature spectrum for several wind velocities calculated from the model [18] is presented in Fig. 6.8. Its characteristic feature is the existence of a maximum in the gravitational-capillary wave region. Gravitational-capillary waves are the main scatterer of the microwave radio waves.

The value of the maximum curvature spectrum nonlinearly changes with increasing wind speed: in low winds the growth is noticeably faster than in strong winds. The similar nature of changes in spectral density level is consistent with direct measurements of the sea surface slope modulus obtained with a laser slopemeter [34].

When probing the sea surface at an angle of 45°, changes in the length of the probing wave from 0.8 cm to 10 cm correspond to changes in the length of the Bregg component in the range from 0.57 cm to 7.1 cm. The corresponding region of the wave number variation is shown in Fig. 6.8 with vertical lines. It can be seen that in this region, as the wind velocity increases, the spectrum of curvature, and therefore the spectrum $\tilde{\Psi}(k)$, grows dozens of times.

Thus, the back-scattered radio signal carries information about the temperature and salinity of seawater. In practice, however, direct measurements of seawater temperature and salinity are not feasible because the relative changes in the spectrum $\Psi\left(\vec{k}_B\right)$ are much greater than the relative changes in the geometric coefficient $\left|G_{pp}(\theta, \varepsilon)\right|^2$ [24].

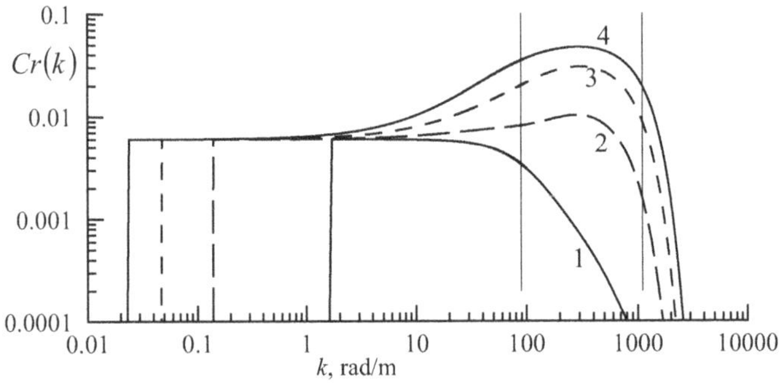

Fig. 6.8 Sea surface curvature spectra $Cr(k)$. Curves 1–4 correspond to wind speeds from 2 to 20 m/s in 6 m/s steps

6.5 Determination of Seawater Temperature and Salinity by Active Ocean Sounding Instruments

Currently, space-based microwave radiometers are the main source of information on the physical and chemical characteristics of the surface layer of the oceans [9, 16]. The output signals of microwave radiometers are proportional to the emissivity of the sea surface, which is small, which imposes limitations on their resolution [1].

Restrictions in the use of radiometers require the development of new measurement methods. One of such methods is the determination of physical and chemical characteristics of the sea surface based on data from active radio-sounding in the microwave range. The radio signal reflected from the sea surface depends on its roughness and relative dielectric permeability of water under the ocean-atmosphere boundary [22, 28, 29]. In turn, the relative dielectric constant depends on the temperature and salinity of the seawater and the length of the sounding radio waves [19]. These dependencies, in principle, make it possible to solve the reverse problem, i.e. to recover temperature and salinity from remotely sensed data.

The problem of measuring the physical and chemical characteristics of seawater by active radio sounding is that the main factor determining the scattered radio signal is the roughness level of the sea surface [2]. Changes in the radio signal roughness caused by changes in the roughness are far superior to those caused by other factors. This fact, in particular, made it possible in 1966 to propose a method for remote determination of the speed of the driving wind by recording backward scattering of radio waves at the ocean-atmosphere boundary [21]. Thus, in order to determine the physical and chemical characteristics of sea water from radar measurements, it is necessary to minimize the influence of sea surface roughness changes.

In order to exclude the dependence of the backscattered radio signal on the sea surface roughness, which is characterized by the spectrum $\Xi\,(\vec{k}_B)$, it was proposed in work [32] to perform sounding of the same surface area on different polarizations. It follows from (6.19) that the polarization ratio does not depend on the spectrum $\Xi\,(\vec{k}_B)$

$$R^0 = \sigma^0_{hh} \,/\, \sigma^0_{vv} = |\, G_{hh}\,(\theta, \varepsilon)\,|^2 / \,|\, G_{vv}\,(\theta, \varepsilon)\,|^2 \tag{6.26}$$

When probing the sea surface in the microwave range, the Bregg components are gravity-capillary and the shortest gravity waves. Since on the sea surface there are always waves whose length is much longer than the length λ_B, the Bregg components spread over the curvilinear surface created by long waves. The curvature of the surface results in a change in the local angle of incidence by the value of the slope angle created by long waves [11, 14 31]. At irradiation of such surface, the local angle of incidence is equal to

$$\theta_L = \theta - \beta_\uparrow \tag{6.27}$$

where $\beta_{\uparrow}$ is angle of the sea surface in the direction of the sensing. As a result of changes in the local angle θ_L, the resonance condition (6.18) changes, i.e., the wavelength at which the scattering occurs changes, and the value of the geometric factor $\left|G_{pp}(\theta_L, \varepsilon)\right|^2$ changes.

The curvature effect of the surface over which the resonant waves propagate can be taken into account by averaging the backscatter cross sections σ_{pp}^0 over the entire angle range $\beta_{\uparrow}$. Averaging is performed with weight proportional to the probability density function $\beta_{\uparrow}$.

$$\sigma_{pp}^L = \int \sigma_{pp}^0(\theta - \beta_{\uparrow})\, P(\beta_{\uparrow})\, d\beta = \int 16\, K^4 \left|G_{pp}(\theta - \beta_{\uparrow})\right|^2 \Xi_r\left(\vec{k}_B\left(K, \theta - \beta_{\uparrow}\right)\right) P(\beta_{\uparrow})\, d\beta_{\uparrow} \quad (6.28)$$

Polarization ratio in the presence of long waves takes the form of

$$R^L = \frac{\int \left|G_{hh}(\theta - \beta_{\uparrow})\right|^2 \Xi_r\left(\vec{k}_B\left(K, \theta - \beta_{\uparrow}\right)\right) P(\beta_{\uparrow})\, d\beta_{\uparrow}}{\int \left|G_{vv}(\theta - \beta_{\uparrow})\right|^2 \Xi_r\left(\vec{k}_B\left(K, \theta - \beta_{\uparrow}\right)\right) P(\beta_{\uparrow})\, d\beta_{\uparrow}} \quad (6.29)$$

where $P(\beta_{\uparrow})$ is a probability density function. It follows from Eqs. (6.26) and (6.29) $R^0 \neq R^L$.

The dependence of the polarization ratio R^L on the level of roughness of the sea surface remains, but becomes significantly less than the similar dependence of the backscattered signal for measurements on a single polarization.

6.6 Variability of Polarization Ratio

Let us estimate how changes in the local incidence angle affect the polarization ratio. Let's assume that the probability density function of angles $\beta_{\uparrow}$ is described by the Gaussian distribution. Calculations will be made for the case when the sea surface is probed in the *Ku* and *L* radio wave bands. In these bands, the radio wavelengths are between 1.6 cm and 2.5 cm and between 15 cm and 30 cm, respectively.

In oceanological studies, as a rule, it is not the slope angles of the sea surface that are determined, but its slopes that are the first spatial derivatives of surface elevations. The gradients of the sea surface are ξ small, which allows for the use of approximation $\beta \approx arctg\, \xi$. The error of this approximation is 5–7%, if the slopes create waves of all scales [33].

The slope variance D_β produced by long waves can be defined using a semi-empirical model [30], which defines the dispersion D_β as part of the total dispersion produced by waves of all scales calculated from optical measurements [6]. When probing along the wind direction, which corresponds to the situation when the influence of long waves is greatest, we get

$$D_\beta = (0.001 + 0.00316\,W)\,(0.3 + 0.02w) \tag{6.30}$$

where W is wind speed (m/s); w is radio frequency (GHz). Average value w for Ku range is 15 GHz, for L range is 1.5 GHz. For calculations of polarization ratio of R^L complex dielectric constant we used the proposed one [19].

In order to evaluate the possibility of using the polarization ratio for temperature and salinity measurements, it is necessary to evaluate the relative R^0 changes in the temperature range from 0° to 30°, as well as the changes in salinity from 0 to 38‰. Calculations were made at wavelengths of cm, $\lambda = 2.1$ cm, 21 cm and 30 cm. Wavelengths of 2.1 cm and 21 cm correspond to approximately the middle of the Ku and L ranges; wavelength 30 cm corresponds to the upper boundary of the L range. The polarization ratio varies greatly when the incidence angle or the radio wavelength changes. Therefore, for convenience of comparison of sensitivity of polarization ratio to changes of temperature and salinity, at measurements in different conditions, rationing is entered so that in the first case at temperature 0° and in the second case at salinity 0‰ the normalized polarization ratio $\tilde{R}^0$ is equal to one.

The results of the calculations characterizing the dependence of the polarization relation on temperature are presented in Fig. 6.9. It can be seen that with increasing incidence angle θ the sensitivity of the parameter $\tilde{R}^0$ to temperature changes increases at all wavelengths. When probing in the centimeter range, the highest sensitivity is observed in the temperature range from 0° to 10°.

Changes in the parameter $\tilde{R}^0$ with changing salinity of sea water are shown in Fig. 6.10. As with the analysis of the effect of temperature changes on the polarization ratio, it can be seen that the sensitivity increases with increasing angle of incidence. There is a clear trend of increasing sensitivity with increasing radio wavelength.

Relative changes of the parameter $\tilde{R}^0$, as can be seen from Fig. 6.2, given in [7], are comparable with relative changes in the coefficient of radiation and its own radio-luminance temperature of a smooth water surface when changing its thermodynamic temperature within the same limits for which Fig. 6.1 is built here.

It has been shown above that the limiting factor in determining temperature and salinity is the presence of surface waves much longer than the length of the sounding radio wave. With the increasing dispersion of the long wavelength slopes, the normalized backscatter cross section increases with both horizontal and vertical polarization probing. But the growth when probing on the horizontal polarization is faster, which leads to a change in the polarization ratio. The influence of this factor is shown in Fig. 6.11, which shows the dependencies of the polarization relation in the presence of long waves $R^L = R^L(W)$, calculated for three radio wavelengths and two angles of incidence.

To estimate the relative changes in the polarization ratio, which occur when the wind speed changes, the normalization $\tilde{R}^L(W) = R^L(W)/\,R^L(W = 1\text{ m/s})$ is introduced. It can be seen that when probing the sea surface in the Ku range, the influence of waves longer than the length of the probing radio waves is higher than in the L range. With increasing angle of incidence, the influence of long waves on polarization ratio $\tilde{R}^L$ increases.

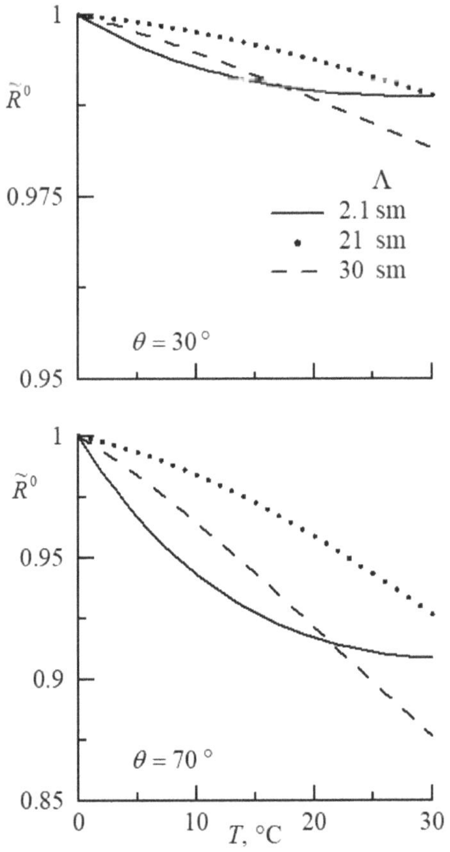

Fig. 6.9 Dependence of the parameter $\tilde{R}^0$ on the temperature T of seawater

It should be noted that the change in sea surface roughness is also a factor that significantly influences radiometric measurements [23] (Meissner and Wentz 2002). The sensitivity of the polarization relation to changes in wind speed has the same order as the sensitivity of the radioluminescence temperature for vertical and horizontal polarizations [7].

In the last decade, active research has been conducted on the dependencies of wind speed on statistical characteristics of slopes created by waves much longer than the length of the probing radio wave [4, 8]. Obtaining these dependencies will make it possible, when determining temperature and salinity from active radio-sounding data, to introduce corrections that take into account changes in sea surface roughness.

Fig. 6.10 Dependence of the parameter $\tilde{R}^0$ on the salinity S of seawater

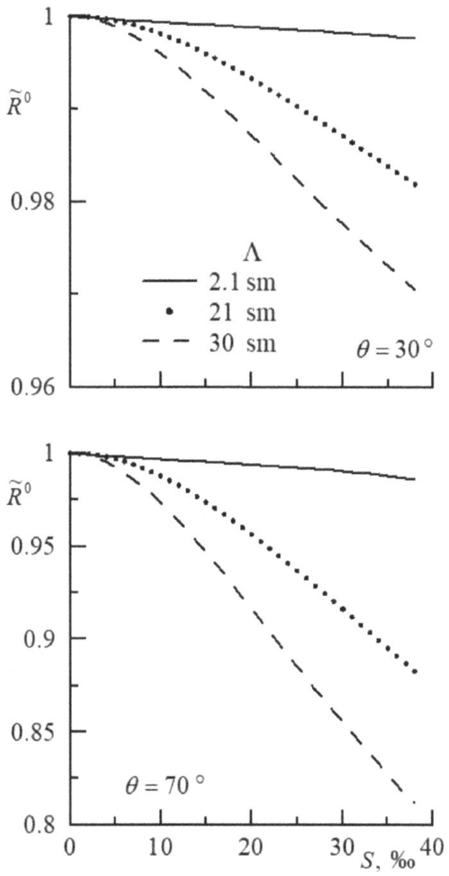

6.7 Conclusion

The known limitations in the use of microwave radiometers for remote determination of seawater temperature and salinity require the development of new approaches to address this problem. One such method is the determination of the physical and chemical characteristics of the sea surface using active microwave radio-sounding data. The radio signal reflected from the sea surface depends on the temperature and salinity of the water under the ocean-atmosphere boundary. These dependencies, in principle, make it possible to solve the reverse problem, i.e. to recover temperature and salinity from remotely sensed data.

The main factor limiting the possibility of direct determination of temperature and salinity from active sensing data is the dependence of the backscattered radio signal on the level of spectral density of resonant surface waves. Changes in the Bregg spectral density caused by changes in the sea surface radio signal are much wider than those caused by variations in temperature or salinity.

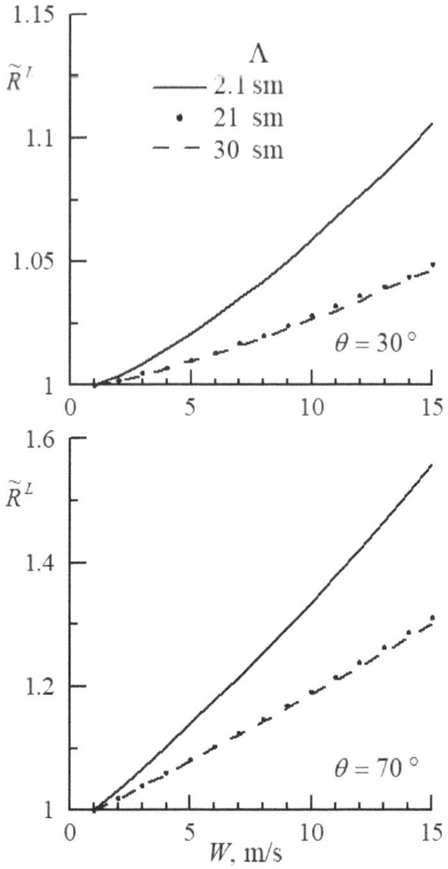

Fig. 6.11 Dependencies of the parameter $\tilde{R}^0$ on wind speed W

When calculating temperature and salinity using the polarization ratio, the effect of changing roughness levels is significantly reduced. It also changes with changing sea surface conditions. The polarization ratio is affected by changes in the local angle of incidence caused by the presence of surface waves much longer than the Breggs wavelength. This effect is roughly the same when measuring the polarization ratio and the radio-luminance temperature.

Active and passive methods for remote determination of sea surface temperature and salinity have the same order of sensitivity. The advantage of passive methods is their small size and lower cost of equipment for their implementation. The advantage of active methods is a significantly higher spatial resolution.

References

1. Armand NA, Tishchenko YG, Savorskiy VP, Smirnov MT, Ablyazov VS, Khaldin AA (2010) Promising space projects with L-band microwave radiometric systems. Issledovaniye Zemli iz kosmosa 1:20–26
2. Bass FG, Fuks IM (1972) Scattering of waves on a statistically uneven surface. Nauka, Moscow, p 424
3. Brown GS (1979) Estimation of surface winds using satelliteborne radar measurements at normal incidence. J Geop Res 84(B8):3974–3978
4. Chen P, Yin Q, Huang P (2015) Effect of non-Gaussian properties of the sea surface on the low-incidence radar backscatter and its inversion in terms of wave spectra by an ocean wave. Chin J Oceanol Limnol 33(5):1142–1156
5. Cheng Y, Liu Y, Xu Q (2006) A new wind-wave spectrum model for deep water. Indian J Mar Sci 35(3):181–194
6. Cox C, Munk W (1954) Measurements of the roughness of the sea surface from photographs of the sun glitter. J Opt Soc Am 44(11):838–850
7. Danilychev MV, Kutuza BG, Kaloshin VA, Moshkov AV (2015) Ispolzovaniye radiometrii SVCh-diapazona dlya izmereniya solenosti poverkhnostnykh vod Mirovogo okeana (Using microwave radiometry to measure the salinity of surface waters of the World Ocean). Zhurnal Radioelektroniki 1:1–13
8. Danilyichev MV, Nikolaev AN, Kutuza BG (2009) The use of Kirchhoff method for practical calculations in microwave radiometry of the rough sea surface. Radiotehnika i elektronika 54(8):915–925
9. Gayfulin DR, Tsyrulnikov MD, Uspensky AB, Kramchaninova EK, Uspensky SA, Svirenko PI, Gorbunov ME (2017) The usage of MTVZA-GYa satellite microwave radiometer observations in the data assimilation system of the Hydrometcenter of Russia. Russ Meteorol Hydrol 42(9):564–573. https://doi.org/10.3103/s1068373917090035
10. Hyatt HA (1970) Emission, reflection, and absorption of microwaves at a smooth air-water interface. J Quant Specrrosc Radim Transfer 10:211–247
11. Janssen PAEM, Wallbrink H, Calkoen CJ, van Halsema D, Oost WA, Snoeij P (1998) VIERS-1 scatterometer model. J Geophys Res 103:7807–7831
12. Kalmykov AI, Pustovoytenko VV (1976) On polarization features of radio signals scattered from the sea surface at small grazing angles. J Geophys Res 81(12):1960–1964. https://doi.org/10.1029/jc081i012p01960
13. Klein LA, Swift CT (1977) An improved model for the dielectric constant of sea water at microwave frequencies. IEEE Trans Ant Prop AP-25:104–111
14. Kudryavtsev V, Akimov D, Johannessen J, Chapron B (2005) On radar imaging of current features: 1. Model and comparison with observations. J Geophys Res 110:C07016. https://doi.org/10.1029/2004jc002505
15. Kudryavtsev V, Hauser D, Caudal G, Chapron B (2002) A semiempirical model of the normalized radar cross-section of the sea surface, 1, Background model. J Geophys Res 107. https://doi.org/10.1029/2001jc001003
16. Kuzmin AV, Repina IA, Sadovskiy IN, Selunskiy AB (2015) Microwave radiometric studies of the sea surface. Sovremennyye probl distantsionnogo zondirovaniya Zemli iz kosmosa 12(5):76–97
17. Kuzmin FV (2016) Impact of variability of physicochemical characteristics of the marine environment on accuracy remote definition of wind speed. J Radio Electr (4). Available at http://jre.cplire.ru/jre/jun19/8/text.pdf
18. Liu Y, Su M-Y, Yan X-H, Liu WT (2000) The mean-square slope of ocean surface waves and its effects on radar backscatter. J Atmos. Oceanic Technol 17:1092–1105
19. Meissner T, Wentz FJ (2004) The complex dielectric constant of pure and sea water from microwave satellite observations. IEEE Trans Geosci Remote Sens 42(9):1836–1849

20. Meissner T, Wentz FJ (2004) The complex dielectric constant of pure and sea water from microwave satellite observations. IEEE Transactions on Geoscience and Remote Sensing 42(9):1836–1849
21. Moore RK, Pierson WJ (1966) Measuring sea state and estimating surface winds from a polar orbiting satellite. In: Proceedings of International symposium on electromagnetic sensing of the earth from Satellites Miami Beach, FL, 22–24 November, pp R1–R28
22. Plant WJ (2002) A stochastic, multiscale model of microwave backscatter from the ocean. J Geophys Res 107(C9):3120. https://doi.org/10.1029/2001JC000909
23. Pospelov MN (2004) Application of polarization radiometry in remote sensing: history and prospects. Sovremennyye probl distantsionnogo zondirovaniya Zemli iz kosmosa 1(1):58–69
24. Sadovsky IN (2008) Methods of remote determining the characteristics of wind waves: 1. Calculation of brightness contrasts of rough water surface. Sovremennyye probl distantsionnogo zondirovaniya Zemli iz kosmosa 2:192–198
25. Sadovsky IN, Kuzmin AV, Sharkov EA, Sazonov DS, Pashinov EV, Asheko AA, Batulin SA (2002) The analysis of dielectric permittivity's models of water environment, used in tasks of remote sensing of water areas, vol 2172. Space Research Institute of RAS, Moscow, 60 p
26. Stogrin A (1971) Equations for calculating the dielectric constant for saline water. JEEE Trans Microw Theor Techn 19(8):733–736
27. Stogryn AP, Bull HT, Rubayi K, Iravanchy S (1995) The microwave dielectric properties of sea and fresh water. GenCorp Aerojet, Azusa, California
28. Terekhin YV, Pustovoitenko VV (1986) Influence of temperature and salinity of sea water on characteristics of microwave radar signal. Earth Obs Remote Sens 2:16–21
29. Valenzuela G (1978) Theories for the interaction of electromagnetic and ocean waves—a review. Bound Layer Meteorol 13(1–4):61–85
30. Wilheit TT (1979) A model for the microwave emissivity of the ocean's surface as a function of wind speed. IEEE Trans Geosci Electron 17(4)
31. Zapevalov AS (2009) Bragg scattering of centimeter electromagnetic radiation from the sea surface: The effect of waves longer than Bragg components. Izv Atmos Ocean Phys 45(2):253–261
32. Zapevalov AS (2018) On definition of temperature and salinity of sea water by means of active sounding of the ocean. Sovremennyye probl distantsionnogo zondirovaniya Zemli iz kosmosa 15(7):131–140. https://doi.org/10.21046/2070-7401-2018-15-7-131-140
33. Zapevalov AS, Lebedev NE (2014) Simulation of statistical characteristics of sea surface during remote optical sensing. Atmos Oceanic Opt 27(6):487–492
34. Zapevalov AS (2002) Statistical characteristics of the moduli of slopes of the sea surface. Phys Oceanogr 12(1):24–31

Chapter 7
Measurements of Statistical Moments of Sea Surface Slope Based on Satellite Sounding Data

7.1 Introduction

Methods for measuring the slopes of the sea surface using spacecraft-based optical instruments based on the registration of reflected sunlight are now widely used [2, 6]. Remote measurements of slope are a class of indirect measurements. The reliability of the data obtained with the help of optical scanners and laser probing depends on the correctness of the models describing the optical signal formation, including the correct description of the sea surface slope distribution [11, 24, 37].

The signal recorded by the optical scanner generates a number of physical mechanisms. These include the reflection of sunlight by the sea surface, mechanisms associated with the scattering of sunlight by the atmosphere, as well as the penetration of sunlight under the atmosphere-ocean boundary and scattering in the water column [16, 32]. The slope of the sea surface are determined under the assumption that the signal recorded on board the satellite is formed as a result of the specular reflection of direct sunlight towards the satellite, the effects associated with the passage of light through the atmosphere are small or can be corrected [8, 15, 45]. The area where the specular reflection mechanism dominates is called sun glint [29]. In this area, the optical scanner signal is determined by the probability density function of slope reflecting sunlight in the direction of the satellite [21].

Studies of sea surface topographic characteristics by optical scanner have limitations [35]. If its signal is formed by sea surface areas with high slope and low probability density values, the influence of the atmosphere increases significantly. Such situations occur when sunlight is reflected from areas of the sea surface away from the sun glint, which leads to errors in determining statistical moments of slopes.

One of the promising optical methods for studying sea surface characteristics is laser sensing [25]. When probing from a high altitude (onboard the satellite) the spot on the sea surface has a diameter of several tens of meters [11]. At the same time, the

A. Zapevalov et al., *Simulation of the Sea Surface for Remote Sensing*, Springer Oceanography,
https://doi.org/10.1007/978-3-030-58752-9_7

laser-lit area contains a large number of elements that meet the condition of specular reflection. Calculations of sea surface characteristics are carried out on the basis of bidirectional reflectance distribution function (BRDF) built for analysis of signals from space-based optical scanners [21]. The general drawback of calculations of statistical estimations of sea surface slope using laser probing data is that they allow to obtain statistical estimations of slopes only within a priori models [43].

7.2 Bidirectional Reflection Distribution Function

The signal recorded on the spacecraft resulting from the reflection of sunlight from the sea surface is related to the statistical distribution of its slope. It is described by the BRDF [31]. This function is determined by the ratio between the brightness of the mirrored surface of solar radiation and the flux density of solar radiation

$$BRDF = \frac{\pi\ I_r(\theta_r, \varphi)}{H_s(\theta_s, \varphi = 0)\ \cos\theta_s} = \frac{\pi\ F_w(\beta)}{4\ \cos\theta_s \cos\theta_v \cos^4\beta} P\left(\xi_x, \xi_y\right)` \qquad (7.1)$$

where $I_r(\theta_r, \varphi)$ is the brightness of the $I_r(\theta_r, \varphi)$ mirror reflection; $H_s(\theta_s, \varphi = 0)$ is solar radiation flux density; θ_s and θ_r are zenith angles of incident solar rays and rays reflected in the direction of the spacecraft; φ is azimuth angle between the directions of incident and reflected rays of light; $F_w(\beta)$ is Fresnel reflection coefficient for sea water; $P\left(\xi_x, \xi_y\right)$ is two-dimensional probability density function of orthogonal components of sea surface slopes ξ_x and ξ_y; β is angle determined by the condition $tg\beta = \sqrt{\xi_x^2 + \xi_y^2}$. Here is the angle $\beta = \frac{1}{2}\arccos(\cos\theta_s \cos\theta_v + \sin\theta_s \sin\theta_v \sin\varphi)$. Mirror reflection of sunlight in the direction of the satellite occurs if two conditions are simultaneously fulfilled for orthogonal components of sea surface slopes [2]

$$\xi_x = -\frac{\sin\theta_s + \sin\theta_r \cos\varphi}{\cos\theta_s + \cos\theta_r} \qquad (7.2)$$

$$\xi_y = -\frac{\sin\theta_r \sin\varphi}{\cos\theta_s + \cos\theta_r} \qquad (7.3)$$

In the analysis of sea surface slope it is common to distinguish two components ξ_c and ξ_u oriented respectively crosswind and upwind direction. The transition from ξ_x and ξ_y to the components ξ_c and ξ_u is done by turning the coordinate system $\xi_u = \xi_x \cos\phi + \xi_y \sin\phi$, $\xi_c = -\xi_x \sin\phi + \xi_y \cos\phi$. The rotation angle ϕ is equal to the angle between the solar azimuth and the wind velocity vector direction [21].

The characteristics of empirical distributions of crosswind ξ_c and upwind ξ_u components of sea surface slope are described in a number of works [2, 4, 12,

19]. It has been found that in the area of large values of slopes there is a considerable variation of empirical estimations of the probability density function of slope components (see, for example, Fig. 4 in [2]). High noise level in these areas was also noted in [6].

Reliable approximations of the slope probability density function can be built only for a range in which the values of the slope component do not exceed some critical threshold ξ_0 [4, 30]. The values of the parameter ξ_0 for transverse and longitudinal slope components meet the following conditions

$$\xi_{0c} < b_0\sqrt{\overline{\xi_c^2}} \tag{7.4}$$

$$\xi_{0u} < b_0\sqrt{\overline{\xi_u^2}} \tag{7.5}$$

where $b_0 \approx 2.5$; $\overline{\xi_c^2}$ и $\overline{\xi_u^2}$ are the variances of the crosswind and upwind component of the slope.

7.3 Restore Statistical Moments from Optical Scanners

The dependence of the optical scanner signal on the two-dimensional distribution of the sea surface slopes allows solving the inverse problem and restoring the slope characteristics using remote sensing data [2, 6, 35, 40].

Nonlinear processes in the sea surface wave field lead to deviations in the slope distribution from the Gaussian distribution [23, 27]. Despite the fact that the deviations are small, they play an important role in the sea surface reflection models of light [43, 34]. The probability density of random quasi-gaussian variables is described using models based on the Gram–Charlier series.

A model describing the two-dimensional probability density function of longitudinal and transverse components of slope based on the Gram–Charlier series was proposed in [4]. This model was named after its authors as the Cox–Munk model.

$$P_{CM}(\xi_c, \xi_u) = P_G(\xi_c, \xi_u)\left[1 - \frac{1}{2}C_{21}H_2\left(\tilde{\xi}_c\right)H_1\left(\tilde{\xi}_u\right) - \frac{1}{6}C_{03}H_3\left(\tilde{\xi}_u\right)\right.$$
$$\left.+\frac{1}{24}C_{40}H_4\left(\tilde{\xi}_c\right) + \frac{1}{4}C_{22}H_2\left(\tilde{\xi}_c\right)H_2\left(\tilde{\xi}_u\right) + \frac{1}{24}C_{04}H_4\left(\tilde{\xi}_u\right)\right] \tag{7.6}$$

where $\tilde{\xi}_c = \xi_c/\sqrt{\overline{\xi_C^2}}$; $\tilde{\xi}_u = \xi_u/\sqrt{\overline{\xi_u^2}}$; C_{ij}—the coefficients of the Gram–Charlier series; H_n—the orthogonal polynomial of the Hermite order n. Here the first index of the coefficient C_{ij} corresponds to the crosswind component of the slope, the second to the upwind one. The Hermite polynomials included in Eq. (7.6) are described by expressions

$$\begin{cases} H_2(x) = x^2 - 1, \\ H_3(x) = x^3 - 3x, \\ H_4(x) = x^4 - 6x^2 + 3. \end{cases} \tag{7.7}$$

In field experiments, as a rule, statistical moments up to and including the fourth order are defined. Therefore, the model $P_{CM}(\xi_c, \xi_u)$ uses truncated series of Gram–Charlier, which includes only the first five members of the series. As a result, the model (7.6) allows to describe the slope distribution only in a limited range of change ξ_c and ξ_u when the conditions (7.4) and (7.5) are met.

Calculation of statistical characteristics of sea surface slope from optical scanners was carried out by selection of seven parameters by minimization of discrepancies between empirical probability densities and Cox–Munk model [2]. The parameters for which the fitting is performed are $\overline{\xi_c^2}, \overline{\xi_u^2}, C_{03}, C_{04}, C_{40}, C_{21}, C_{22}$. It follows from the conditions (7.4) and (7.5) that the mentioned procedure can be performed only within the range where the model $P_{CM}(\xi_c, \xi_u)$ is defined and in the range where the empirical function of the probability density of the sea surface slopes is constructed using optical images. Thus, the model fitting can be carried out only in the limited range of slope changes, which leads to the calculation errors of their statistical moments. Obviously, the values of skewness and kurtosis of slope components obtained from truncated distributions depend on the boundary points of these distributions.

Let's build the skewness and kurtosis dependencies on the boundaries in which the probability density function of the upwind component of slopes is defined. The field of surface waves is weakly non-linear. This results in deviations from the Gaussian distribution. Since the nonlinear effects are small, the deviations are not large, and the mixed coefficients are small, which is confirmed by measurement data [2, 4, 19]. Let us suppose that the mixed moments can be neglected, and the model of the distribution of the upwind component of the slope can be presented as follows

$$\widetilde{P}_{CM}\left(\tilde{\xi}_u\right) = P_G\left(\tilde{\xi}_u\right)\left[1 - \frac{C_{03}}{6} H_3\left(\tilde{\xi}_u\right) + \frac{C_{04}}{24} H_4\left(\tilde{\xi}_u\right)\right] \tag{7.8}$$

Provided that $\overline{\xi}_u = 0$ the statistical moments of the order n for $\widetilde{P}_{CM}\left(\tilde{\xi}_u\right)$ equal

$$\mu_{CMn} = \int_{-\infty}^{\infty} \tilde{\xi}_u^n \, \widetilde{P}_{CM}\left(\tilde{\xi}_u\right) d\tilde{\xi}_u \tag{7.9}$$

where n—the order of the statistical moment.

Hermit's polynomials have the property of orthogonality.

$$\int_{-\infty}^{\infty} H_n(x) \cdot H_m(x) \cdot PN(x)\,dx = \begin{cases} 0 & npu\ n \neq m \\ n!, & npu\ n = m \end{cases} \tag{7.10}$$

where $PN(x) = \frac{1}{\sqrt{2\pi}} \exp\left(-\frac{1}{2}x^2\right)$. Taking into account (7.10), if we formally calculate the moments μ_{CM3} and a for $\tilde{\xi}_u$, integrating within the range from $-\infty$ to ∞, without paying attention to negative values $\widetilde{P}_{CM}\left(\tilde{\xi}_u\right)$, we get $C_{03} \equiv \mu_{CM3}$ and $C_{04} \equiv \mu_{CM4} - 3$. The mentioned procedure is not correct because the model $P_{CM}(\xi_c, \xi_u)$ is defined within the limits of the given (7.4) and (7.5).

Change the integration limits in (7.9), bringing them closer to b_0. Consider the dependencies

$$\begin{cases} \mu_{CM3} = \mu_{CM3}(b),\ C_{04} = 0 \\ \mu_{CM4} = \mu_{CM4}(b),\ C_{03} = 0 \end{cases} \tag{7.11}$$

where b—is a symmetrical integration limit in (7.9). The results of the calculations shown in Fig. 7.1. show that the existing limitations of the Cox–Munk model have a significant impact on the possibility of restoring statistical moments of sea surface slopes using optical scanners. The strong dependence of even statistical moments when the values of parameter b change near the value b_0 also takes place for Gaussian

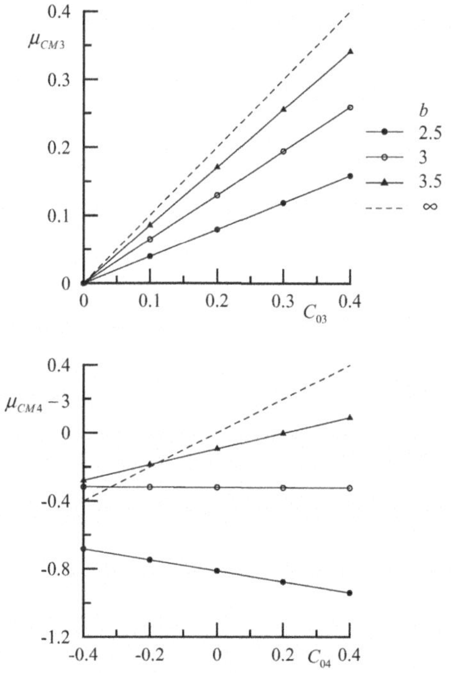

Fig. 7.1 Dependence of statistical distribution moments (7.8) on integration limits

distribution. If the parameter b takes values 2.5, 3 and 3.5, the kurtosis of the truncated Gaussian distribution $\mu_{CM4} - 3$ is −0.81, −0.32 and −0.09 [35].

In order to extend the integration limits and to bring the coefficients C_{03} and C_{04} closer to the statistical moments of slope, we extrapolate the probability density function of slope outside the region (7.5). For this purpose, we use the combined model, which in the region (7.5) is close to the Gram–Charlier distribution, and outside it is close to the Gaussian distribution [39]. The combined model of the upwind component of slope has the form

$$\widetilde{P}_C\left(\tilde{\xi}_u\right) = P_G\left(\tilde{\xi}_u\right)\left\{1 + \left[-\frac{C_{03}}{6}H_3\left(\tilde{\xi}_u\right) + \frac{C_{04}}{24}H_4\left(\tilde{\xi}_u\right)\right]F\left(\tilde{\xi}_u\right)\right\} \tag{7.12}$$

where function F acts as a filter.

A two-parameter function can be used as a filter

$$F\left(\tilde{\xi}_u\right) = \exp\left[-\left(\left|\tilde{\xi}_u\right|/d\right)^m\right] \tag{7.13}$$

where the parameter d determines the area inside which $\widetilde{P}_C\left(\tilde{\xi}_u\right) \approx \widetilde{P}_{CM}\left(\tilde{\xi}_u\right)$; the parameter m determines the speed at which the model $\widetilde{P}_C\left(\tilde{\xi}_x\right)$ approaches the Gaussian distribution outside the area (7.5).

The analysis of sea surface roughness obtained in different experiments showed its strong variability [17] Lidar; [41, 34]. The values of the parameter C_{03} depend on the wind speed and mainly lie in the range from 0 to −0.4. The values of the parameter C_{04} depend on the wind speed, according to measurements by a laser slopemeter $C_{04} = 0.43 \pm 0.46$.

One-dimensional models $\widetilde{P}_{CM}\left(\tilde{\xi}_u\right)$ and $\widetilde{P}_C\left(\tilde{\xi}_u\right)$, built at parameter values $C_{03} = -0.3$, $C_{04} = 0.2$ are shown in Fig. 7.2. The parameters of the filter $F\left(\tilde{\xi}_u\right)$ are selected equal $d = 3.5$ and $m = 3$.

To estimate the difference between the models outside the area (7.5), where the probability density function is small, their relations are shown in Fig. 7.1. In the area (7.5) the models $\widetilde{P}_{CM}\left(\tilde{\xi}_u\right)$ and $\widetilde{P}_C\left(\tilde{\xi}_u\right)$ are practically identical, outside this area at the $\tilde{\xi}_u < -3$—the model $\widetilde{P}_{CM}(\xi_u)$ has negative values [35].

Let us consider how changes in the integration limit b of the range for which statistical moments are calculated affect their magnitude (Fig. 7.3). For calculations we use the model $\widetilde{P}_C\left(\tilde{\xi}_u\right)$

$$\mu_{Cn}(b) = \int_{-b}^{b} \tilde{\xi}_u^n \widetilde{P}_C\left(\tilde{\xi}_u\right) d\tilde{\xi}_u \tag{7.14}$$

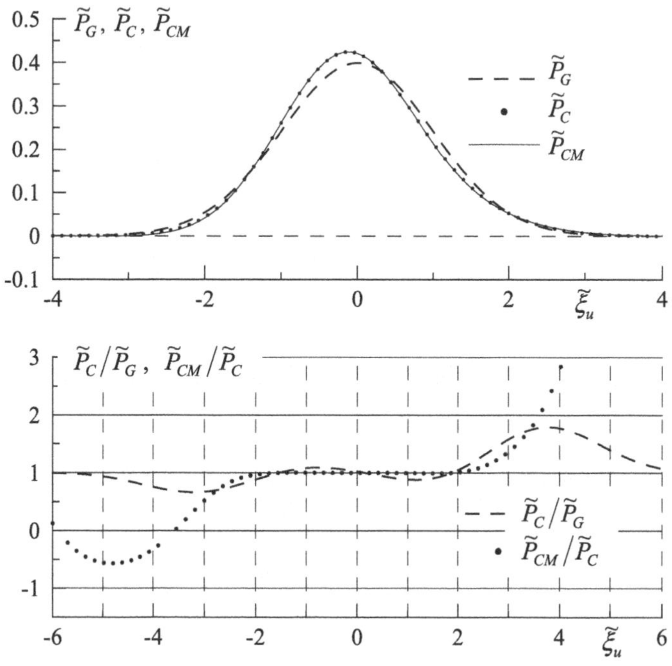

Fig. 7.2 Probability density functions of the longitudinal component of the slope of the sea surface

Fig. 7.3 Dependence of statistical moments of the third μ_{C3} and fourth μ_{C4} orders of magnitude on integration limit b, calculated for the combined model

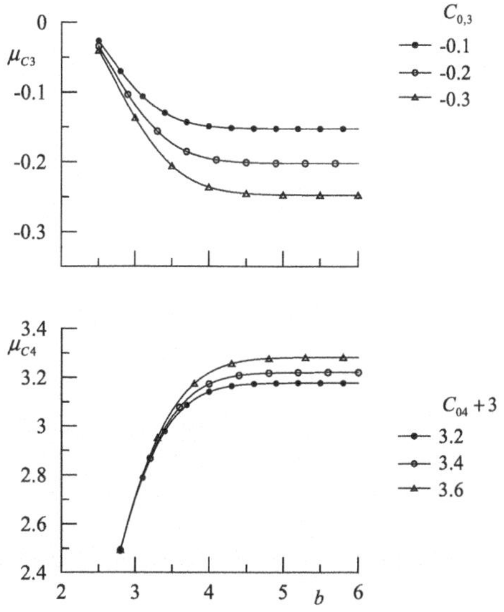

Table 7.1 Model parameters $\widetilde{P}_C\left(\tilde{\xi}\right)$ and statistical moments calculated from it

No.	C_{03}	μ_{C3} for $b = 6$
1	−0,1	−0,034
2	−0.2	−0.06
3	−0.3	−0.10
No.	$C_{04} + 3$	μ_{C4} for $b = 6$
1	3.2	3.008
2	3.4	3.016
3	3.6	3.024

The numerical values of the parameters C_{03} and C_{04}, the model $\widetilde{P}_C\left(\tilde{\xi}\right)$ and the values μ_{C3} and μ_{C4} calculated for it at $b = 6$ are given in Table 7.1.

The results of numerical modeling within the limits of changes of parameters C_{03} and C_{04} observed in experiments allowed to draw the following conclusion. In order to obtain estimates of statistical moments in the area where they slightly change with changes in the limits of integration, in skewness calculations it is necessary that the condition is met $b > 3.5$, in the calculations of the kurtosis it is necessary that the condition $b > 4.5$ is met.

The dependencies presented in Fig. 7.3 $\mu_{C3} = \mu_{C3}(b)$ and $\mu_{C4} = \mu_{C4}(b)$ show that the choice of the function extrapolating the model $\widetilde{P}_{CM}\left(\tilde{\xi}_u\right)$ beyond the region (7.5) critically affects the values of the third and fourth statistical moments. Although the combined model improves the situation, the problem is not completely solved. It is necessary to create a fairly simple analytical model that would allow more accurate restoration of the senior statistical moments of slopes, according to remote sensing data [44].

7.4 Laser Sensing of the Sea Surface from a Satellite

The research of sea surface characteristics by laser sensing can be divided into two directions. The first direction corresponds to the conditions in which probing is carried out from a small height (about 10 m), and the laser spot on the surface has a diameter of several millimeters [41]. In the area of the laser spot periodically there are areas of the sea surface, the slope of which meets the condition of the laser beam reflection in the aperture of the photoreceiver. The signal at the output of the lidar photoreceiver represents a sequence of pulses of different duration and amplitude, which allow to investigate topographic inhomogeneities on the rough sea surface [18].

When probing from high altitude (from an aircraft or spacecraft), the spot on the sea surface has a diameter of several tens of meters [11]. The area illuminated by laser contains a large number of elements that meet the condition of mirror reflection. Here we will limit ourselves to considering this situation.

The task of determining statistical moments of sea surface slopes using laser-sounding data is similar to the task of determining them using radio-sounding data, which has received much attention in recent years [10, 13, 14]. At the same time, there is a fundamental difference [36]. The radio signal reflected from the sea surface forms sea waves much longer than the radio wavelength. Since radio-sounding, as a rule, is carried out in the centimeter range, the obtained estimates correspond to the distribution of slopes created by the waves with the length greater than 10 cm.

Analysis of the laser-sensed sea surface data from spacecraft is usually performed within the framework of a continuous source model [7]. This allows the recorded signal to be described using the BRDF optical images used in the analysis. If the sensing is monostatic, i.e., the source and receiver of the signal are at the same point, the conditions of the mirror reflection are $\theta_s = \theta_r$ $\varphi = \pi$. If the lidar beam is directed vertically downwards, then $\theta_s = \theta_r = 0$.

Using the function (1) assumes that the duration of the laser pulse is sufficiently long, and in some time interval the simultaneous contribution to the reflected signal is given by all elements of the surface that meet the condition of mirror reflection. During the pulsed probing it is necessary to take into account that the mirror reflection points are located at different heights. If the pulse is short, some of the mirror reflection points may not fall into the area currently illuminated by the probing pulse. The length of the pulse in the direction L of probing, is determined by its duration τ, $L = c\tau$, where c is the speed of light. If the length of the pulse L is equal to the height h of the wave, all points of the mirror reflection on its surface will be simultaneously irradiated, however, the photoreceiver will receive the light reflected from the trough and from the crest with a time delay $\Delta t = h/c$. In order to simultaneously fix the reflection from all elements located on the wave height h, it is necessary to fulfill the condition $\tau > 2h/c$ [38].

As a parameter characterizing the height of the waves, it is customary to use a significant wave height h_s. The significant wave height is equal to the average value of the height of one third of the highest waves [28]. It should be taken into account that in the wave field there are waves whose heights exceed the height h_s. Relatively rarely observed extreme waves whose criterion is a twofold excess of a significant wave height are excluded [20]. Thus, in order to simultaneously fix reflection from all surface elements, the following condition must be met

$$\tau > a2h_s/c \tag{7.15}$$

where a is dimensionless parameter The value of the parameter a is defined below.

Let us consider a simple model on the basis of which we analyze the dependence of the amplitude of the reflected pulse on the duration of the probing. Let us assume that the vertically directed probing pulse of rectangular shape is reflected from the surface whose elevations are evenly distributed in the range from 0 to z_0. Let us suppose that the density of probability of mirror points distribution by height coincides with the density of probability of elevation on the sea surface. This assumption is used in the analysis of radio-altimeter measurements data [26]. In this case the probability density function of the mirror points distribution at height is described by the expression

$$P_R(z) = \begin{cases} 0, & z < 0 \\ 1, z_0, & 0 \le z \le z_0 \\ 0, & z > z_0 \end{cases} \tag{7.16}$$

To compare the forms of reflected pulses of different duration, let us introduce normalization

$$\widetilde{I}_R(t,\tau) = I_R(t,\tau)/\max(I_R(t,\tau \ge \tau_0)) \tag{7.17}$$

where $\tau_0 = 2z_0/c$, $\max(I(t,\tau \ge \tau_0))$—the maximum value of the reflected pulse, if the duration of the probing pulse is greater or equal τ_0. Here, the index "R" means that the characteristics of the pulse are obtained for the distribution of the mirror points by height in the form (7.16).

The forms of the reflected impulses $\widetilde{I}_R(t)$, constructed at different durations of the probing impulse, are presented in Fig. 7.4. At $\tau < \tau_0$ the impulse $\widetilde{I}_R(t)$ has a trapezoidal form, its maximum value increases with increasing τ. At $\tau = \tau_0$ the impulse $\widetilde{I}_R(t)$ gets a triangular form. Further increase does τ not lead to the growth of the maximum value $\widetilde{I}_R(t)$, and its form again becomes trapezoidal. The forms of reflected pulses are determined by the ratio between the duration of the probing pulse and the parameter τ.

Let us assume that the deviations of the sea surface elevation distributions from the Gaussian distribution are small [42]. Let us use a linear model of the sea surface wave field, in which the wave field is represented as a sum of a large number of independent sinusoidal components. The distribution of the mirror reflection points by height is described by the expression

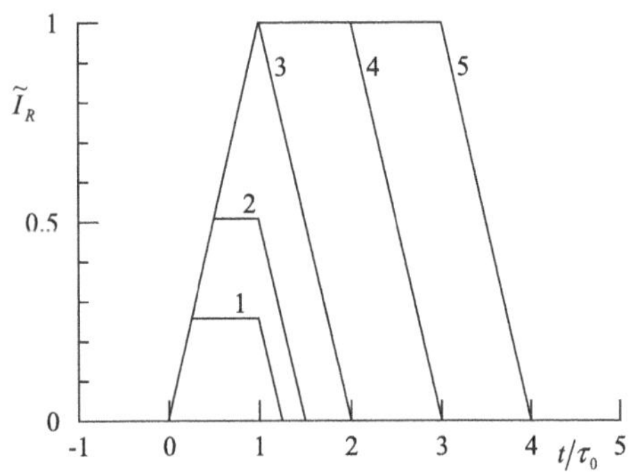

Fig. 7.4 The form of reflected pulses $\widetilde{I}_R(t,\tau)$ calculated for the sea surface, the distribution of the elevations of which is described by the expression (7.16)

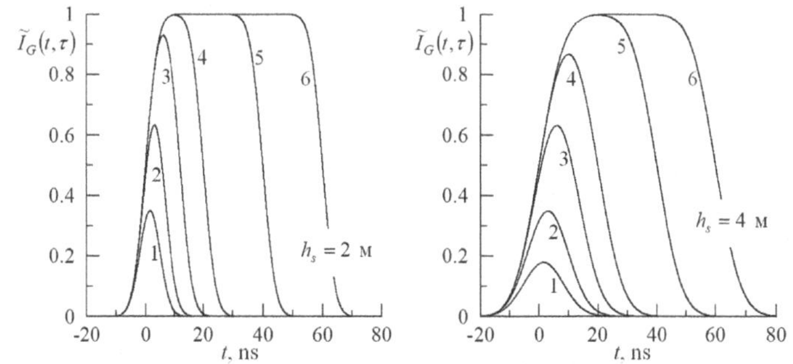

Fig. 7.5 Reflected pulses $\widetilde{I}_G(t, \tau)$ calculated for the sea surface whose elevation distribution is described by the distribution (7.18). The curves 1–6 correspond to the probing pulse duration 3, 6, 12, 20, 40, 60 ns

$$P_G(z) = \frac{1}{\sqrt{2\pi}\sigma_z} \exp\left(-\frac{z^2}{2\sigma_z^2}\right) \tag{7.18}$$

where σ_z^2—variance of the sea surface elevations. Hereinafter, the index "G" means that the characteristics are obtained for the distribution of the mirror points in height in the form (7.18). The parameters h_s are σ_z related by the relation $h_s = 4\sigma_z$ [28].

The forms of the reflected impulses at different durations of the probing impulse and two values of the significant height for the case, when the distribution of the mirror points corresponds to the function (7.18), are shown in the Fig. 7.5. Here, as well as at the construction of Fig. 7.4, the normalization of amplitudes of the reflected impulses according to (7.17) is used.

Qualitative changes in amplitude $\widetilde{I}_G(t, \tau)$ and $\widetilde{I}_R(t, \tau)$ with increasing duration of the probing pulse are similar. In the area of small durations, the amplitude of reflected pulses grows with their growth. At some value τ, which depends on a significant wave height, the amplitudes of reflected pulses reach their maximum, and further increase τ practically does not affect the amplitude.

Changes in the amplitude of the reflected pulse are determined by the ratio between the duration of the probing pulse and the characteristic time scale T, which depends on the vertical distribution of mirror points. For the height distribution of the mirror points in the form (7.16), the characteristic time scale is equal to

$$\mathrm{T}_R = 2z_0/c \tag{7.19}$$

for the distribution (7.18) it can be taken equal.

$$\mathrm{T}_G = 2h_s/c \tag{7.20}$$

The dependences of pulse amplitudes $\widetilde{I}_R$ and $\widetilde{I}_G$ on dimensionless parameters τ/T_R and τ/T_G are presented in Fig. 7.6. The amplitude of the impulse $\widetilde{I}_R$ linearly

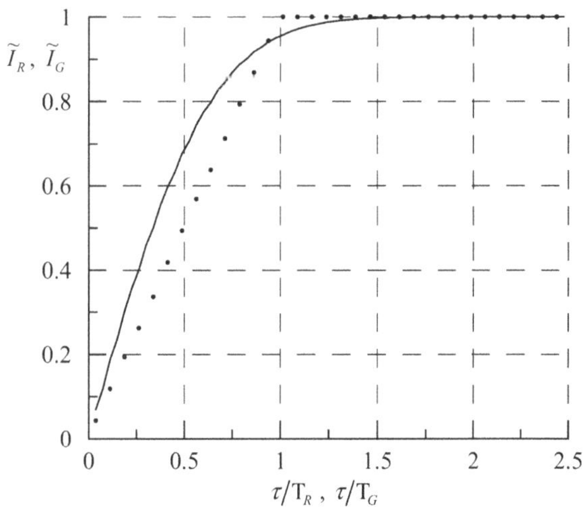

Fig. 7.6 Dependence of the amplitude of reflected pulses $\widetilde{I}_R$ and $\widetilde{I}_G$ from dimensionless parameters τ/T_R and τ/T_R: dashed line is $\widetilde{I}_R = \widetilde{I}_R(\tau/T_R)$; continuous line is $\widetilde{I}_G = \widetilde{I}_G(\tau/T_R)$

grows with growth τ/T_R, reaching units at $\tau/T_R = 1$. The dependence of the pulse amplitude $\widetilde{I}_G$ from τ/T_G is nonlinear. At $\tau/T_G = 1$ we receive $\max\widetilde{I}_G = 0.95$. Returning to the expression (7.15), it is possible to note, that in a case when distribution of heights of points of mirror reflection is described by distribution (7.18), at $a = 1$ possible to fix simultaneously a signal of 95% of points of mirror reflection. Thus, in order to measure the slope variance with a relative error of at least 5%, it is necessary that the duration of the probing pulse exceeds the characteristic time scale T_G.

Expression (7.20) restricts the use of pulsed lidars to determine sea surface roughness. In 2006, the CALIPSO satellite with CALIOP (Cloud-Aerosol Lidar with Orthogonal Polarization) was launched to study atmospheric aerosol and cloudiness [3]. The capabilities of the equipment installed on the satellite have made it possible to expand the range of tasks solved by the satellite and to conduct studies of the relationship between the sea surface slope variance and wind speed [11]. The length of pulses generated by CALIOP lidar is 20 ns, which corresponds to $h_s =$ 3 m.

From the data of vertical laser sounding it is impossible to determine the senior slope semi-invariants as well as wave field anisotropy. It is assumed that the older semi-invariants are small and calculations are performed for isotropic Gaussian distribution [11]. Let us estimate the errors of determining the sea surface slope variance that occur in the calculations as a result of this assumption.

The analysis will start with errors caused by deviations of sea surface slopes from Gaussian distribution. With vertical laser sounding, the reflection in the direction of the spacecraft creates facets whose orientation is close to the horizontal. It can be assumed that $\xi_u = \xi_c = 0$. Polynomials H_1 and H_3 are odd functions, so at small values of the arguments (ξ_u and ξ_c) the contribution of members proportional to these

Table 7.2 Estimates of coefficients C_{22}, C_{04} and C_{40}

References [4]	References [2]	References [19]
0.12 ± 0.06	0.12 ± 0.03	0.17 ± 0.27
0.23 ± 0.41	0.4 ± 0.1	0.43 ± 0.46
0.40 ± 0.23	0.3 ± 0.05	0.33 ± 0.43

polynomials can be neglected. Polynomials H_2 and H_4 are even functions. Taking into account that $H_2(0) = -1$, $H_4(0) = 3$ for $\xi_u = \xi_c = 0$, we get

$$P(\xi_c = 0, \xi_u = 0) = \frac{1}{2\pi\sqrt{\overline{\xi_u^2}\,\overline{\xi_u^2}}}\left[1 + \frac{C_{22}}{4} + \frac{C_{04} + C_{40}}{8}\right] \tag{7.21}$$

The error caused by non-linearity of surface waves is determined by the multiplier in square brackets

$$\Delta_{NG} = \frac{C_{22}}{4} + \frac{C_{04} + C_{40}}{8} \tag{7.22}$$

Studies of the sea surface slope variability carried out with different types of equipment have shown that the coefficients C_{22}, C_{04} and do C_{40} not depend on wind speed. Let us consider three groups of coefficients C_{22}, C_{04} and C_{40} those obtained as a result of processing aerial photographs [4], data processing of optical scanners installed on spacecraft [2], in experiments carried out in marine conditions with the help of laser slope meters on an oceanographic platform [19]. These data are presented in Table 7.2.

In all experiments, the average coefficient values C_{22}, C_{04} and C_{40} are positive. The values of the probability density function at the point $\xi_c = 0$, $\xi_u = 0$ obtained for the same slope variance values are higher than for the Gaussian distribution. This means that the calculations of the slope variance obtained from laser sounding data, which do not take into account the nonlinearity of surface waves, have a systematic error. This error leads to an underestimation of the slope variance. Error value Δ_{NG} for those given in Table 7.2. the average values of the coefficients C_{ij} are between 11 and 14% [25].

The angular energy distribution of sea surface waves is not isotropic [5, 9, 33]. The variance of slopes oriented in the wind direction is higher than the variance in transverse direction. As a parameter characterizing the anisotropy of slope, a three-dimensional index is usually used, which is equal to the ratio of mean-square values of slope in crosswind $\sqrt{\overline{\xi_c^2}}$ and upwind $\sqrt{\overline{\xi_u^2}}$ directions [22].

$$\gamma = \sqrt{\overline{\xi_c^2}} / \sqrt{\overline{\xi_u^2}} \tag{7.23}$$

For waves with long crests the parameter is γ close to zero, for waves with short crests it γ is close to one.

In a linear isotropic wave field, the two-dimensional probability density function of slopes ξ_u and ξ_c is described by the two-dimensional Gaussian distribution

$$P_I(\xi_u, \xi_c) = \frac{1}{2\pi\sqrt{\overline{\xi_u^2}\,\overline{\xi_u^2}}} \cdot \exp\left[-\frac{1}{2} \cdot \left(\frac{\xi_u^2}{\overline{\xi_u^2}} + \frac{\xi_c^2}{\overline{\xi_c^2}}\right)\right] \quad (7.24)$$

Anisotropy of slope affects the characteristics of the reflected signal in laser probing. In expression (7.22), let's move from Cartesian to polar coordinate system. To make this transition, let us present two orthogonal components of the slope as follows

$$\xi_c = \xi_m \sin\varphi, \quad \xi_u = \xi_m \cos\varphi \quad (7.25)$$

where $\xi_m = \sqrt{\xi_u^2 + \xi_c^2}$—slope module; $\varphi = \arctan(\xi_c/\xi_u)$—azimuth angle.

For the probability density function, the condition of transition to new variables

$$P(\xi_m, \varphi) = \frac{\partial\,(\xi_c, \xi_u)}{\partial\,(\xi_m, \varphi)} P(\xi_c, \xi_u) \quad (7.26)$$

where $\frac{\partial\,(\xi_c, \xi_u)}{\partial\,(\xi_m, \varphi)} = \xi_m$—Jacobian.

In the polar coordinate system, the expression (7.22) looks like

$$P_I(\xi_m \varphi) = \frac{\xi_m}{2\pi\sqrt{\overline{\xi_u^2}\,\overline{\xi_c^2}}} \cdot \exp\left[-\frac{1}{2} \cdot \left(\frac{\xi_m^2 \cdot \cos^2\varphi}{\overline{\xi_u^2}} + \frac{\xi_m^2 \cdot \sin^2\varphi}{\overline{\xi_c^2}}\right)\right] \quad (7.27)$$

or

$$P_I(\xi_m \varphi) = \frac{\xi_m}{2\pi\,\gamma^2\overline{\xi_u^2}} \cdot \exp\left[-\frac{1}{2\overline{\xi_u^2}} \cdot \left(\xi_m^2 \cdot \cos^2\varphi + \frac{\xi_m^2 \cdot \sin^2\varphi}{\gamma^2}\right)\right] \quad (7.28)$$

It follows from (7.26) that the slope module distribution depends on the three-dimensional index γ.

Let us exclude in (7.25) the dependence on the azimuth angle, having integrated $P(\xi_m, \varphi)$ in all directions

$$P_m(\xi_m) = \int_0^{2\pi} P(\xi_m, \varphi)\, d\varphi \quad (7.29)$$

In the linear anisotropic wave field at $\gamma = 1$, the probability density function of the slope module ξ_m is described by the Relay distribution [40]. Let us assume

$\overline{\xi_u^2} = \overline{\xi_c^2} = \overline{\xi_A^2}$ that then the distribution looks like:

$$P_A(\xi_m) = \frac{\xi_m}{\sqrt{\overline{\xi_A^2}}} \cdot \exp\left[-\frac{1}{2} \cdot \frac{\xi_m^2}{\overline{\xi_A^2}}\right] \quad (7.30)$$

At vertical probing the laser beam reflected from the sea surface falls into the aperture of the photoreceiver if the slope module ξ_m does not exceed some critical value [1]. The amplitude of the signal recorded on a spacecraft is directly proportional to the probability that

$$\xi_m \le \xi_{m0} \quad (7.31)$$

Introduce rationing

$$\tilde{\xi}_m = \frac{\xi_m}{\sqrt{(1+\gamma^2)\,\overline{\xi_u^2}}} \quad (7.32)$$

Since the variance of slope depends on wind speed, different values $\tilde{\xi}_{m0}$ correspond to the same value ξ_{m0} at different wind speeds [25]. Changes in the parameter $\tilde{\xi}_{m0}$ with increasing wind speed W are shown in Fig. 7.7. The dependences of the slope variance obtained in [2] were used in calculations.

Let us estimate the errors of determining the slope variance caused by the surface wave field anisotropy not taken into account during vertical laser sounding. For this purpose, let us enter the parameter

$$\varepsilon(\gamma) = F_G(\gamma)/F_G(\gamma = 1) \quad (7.33)$$

where F—the probability that the slope module ξ_m meets the condition (7.29).

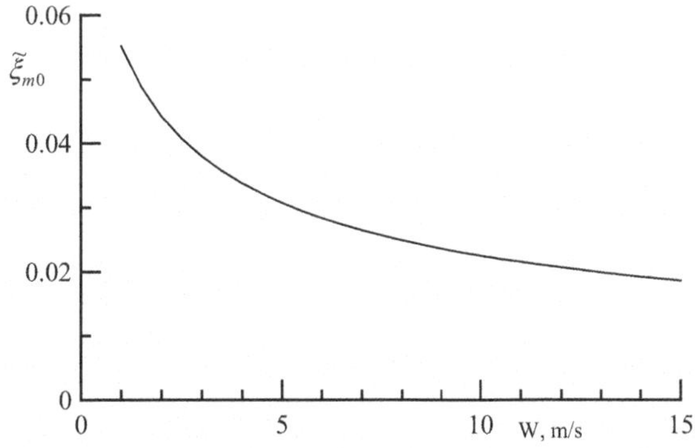

Fig. 7.7 The dependence of the parameter $\tilde{\xi}_{m0}$ on the wind speed W. The calculation of the parameter $\tilde{\xi}_{m0}$ was carried out for the situation corresponding to the conditions of sea surface probing by CALIOP lidar [11]

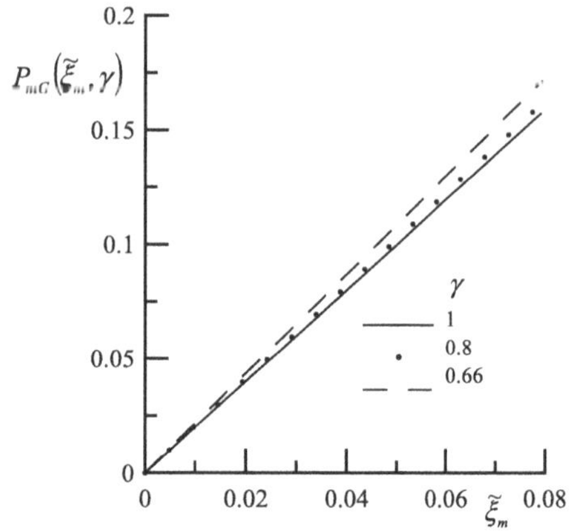

Fig. 7.8 Changes in the probability density function of the module $P_{mG}\left(\tilde{\xi}_m, \gamma\right)$ in the range of small values of the parameter $\tilde{\xi}_m$

In the area of small values of the slope module $\tilde{\xi}_{m0}$, as shown in Fig. 7.8, its probability density function can be approximated by the linear dependence of the slope angle of this approximation changes with the change of the parameter, γ respectively, the probability F changes.

According to the measurement data obtained with a laser inclinometer [19], showed that the average value of the three-dimensional index $\gamma = 0.8$ parameter $\varepsilon = 1.025$. Approximate values of the three-dimensionality index were obtained using the measurement data of optical scanners [2].

Thus, the errors caused by anisotropy of surface wave fields are several percent. These can be neglected in comparison with errors caused by deviations in the slope distribution from the Gaussian distribution.

7.5 Conclusion

The limitations of determining statistical moments of sea surface slope using optical scanners and lasers installed on spacecraft have been analyzed. Both methods are indirect, and the accuracy of measurements is determined by the correctness of our ideas about the optical signal formation.

When using optical scanners, the calculation of statistical slope moments is carried out by calculating the BRDF model parameters from sea surface images. The accuracy limitations are caused by two factors. The first factor is BRDF errors in situations where sunlight in the direction of the spacecraft reflects surface elements whose slopes have low probability densities. However, the physical mechanisms associated with the scattering of sunlight by the atmosphere as well as the penetration of sunlight under the atmosphere-ocean boundary and scattering in the water column make a significant contribution to optical imaging. The condition that an optical image is formed only as a result of mirror reflection ceases to be fulfilled. As a result, the empirical density of probabilities is determined in the limited range of sea surface slope changes, which leads to errors in calculating their statistical moments.

The second factor is related to the selection of the BRDF, which includes a model of the probability density function of slope of the sea surface. For the nonlinear wave field, which is the sea-wave field, the probability density of slopes is approximated by the two-dimensional Cox–Munk model based on the truncated Gram–Charlier series. Calculation of statistical characteristics of sea surface slopes is carried out by selection of Cox–Munk model parameters by minimizing its difference with the probability density function built on optical images. Cox–Munk model allows to correctly describe the probability density function of slopes in a limited range. Outside this range, Cox–Munk model distorts the slope distribution, and negative values appear. Thus, the calculation of statistical moments of slopes using Cox–Munk model corresponds to the calculation of truncated distribution, which is the source of errors.

With vertical laser probing from a satellite, the BRDF degenerates. Errors are caused by deviations of the real wave field from the Gaussian isotropic surface. Uncounted deviations of slope distributions from the Gaussian distribution result in a systematic error of 11–14%. The calculated slope variance values are underestimated. Anisotropy of sea surface slopes, which cannot be determined with vertical probing, leads to an error, which also systematically underestimates the variance values. This error is on average less than 3%.

Influence of duration of a probing pulse on accuracy of definition of variance of slopes is considered. It is shown that for correct determination of variance of slopes it is necessary that the duration of probing impulses exceeds some characteristic time scale, which depends on the significant wave height. If the mirror points are distributed according to the Gaussian law, then to measure the slope variance with a relative error of less than 5%, it is necessary that the duration of the probing pulse exceeds the time scale $T_G = 2h_s/c$.

References

1. Aleksandrov AP, Legeza VP (1988) Laser meter of rough sea surface rms slopes. Physical Oceanography 6:51–56
2. Bréon FM, Henriot N (2006) Spaceborne observations of ocean glint reflectance and modeling of wave slope distributions. J Geoph Res Oceans 111(C6)
3. Chand D, Anderson TL, Wood R, Charlson RJ, Hu Y, Liu Z, Vaughan M (2008) Quantifying above-cloud aerosol using spaceborne lidar for improved understanding of cloudy-sky direct climate forcing. J Geophys Res 113:D13206
4. Cox C, Munk W (1954) Measurements of the roughness of the sea surface from photographs of the sun glitter. J Opt Soc Am 44(11):838–850
5. Donelan MA, Hamilton J, Hui WH (1985) Directional spectra of wind-generated waves. Philos Trans Roy Soc A315:509–562
6. Ebuchi N, Kizu S (2002) Probability distribution of surface wave slope derived using Sun glitter images from geostationary meteorological satellite and surface vector winds from scatterometers. J Oceanogr 58:477–486

7. Ginneken B, Stavridi M, Koenderink J (1998) Diffuse and specular reflectance from rough surface. Appl Opt 37:130–139
8. Ginzburg AS, Melnikova IN, Samulenkov DA, Sapunov MV, Katkovsky LV (2016) Simple optical model of clear sky and cloudy atmosphere for calculation of solar irradiance. Sovremennye problemy distantsionnogo zondirovaniya Zemli iz kosmosa 13(2):132–149
9. Hasselmann DE, Dunckel M, Ewing JA (1980) Directional wave spectra observed during JONSWAP 1973. J Phys Octanogr 10(8):1264–1280
10. Hou Y, Song G, Zhao X et al (2006) Statistical distribution of nonlinear random water wave surface elevation. Chin J Oceanol Limnol 24(1):1–5
11. Hu Y, Stamnes K, Vaughan M, Pelon J, Weimer C, Wu D, Cisewski M, Sun W, Yang P, Lin B, Omar A, Flittner D, Hostetler C, Trepte C, Winker D, Gibson G, Santa-Maria M (2008) Sea surface wind speed estimation from space-based lidar measurements. Atmos Chem Phys 8:3593–3601
12. Hughes BA, Grant HL, Chappell RWA (1977) A fast response surface-wave slope meter and measured wind-wave components. Deep-Sea Res 24(12):1211–1223
13. Karaev V, Kanevsky M, Cotton D, Challenor P (2002) Is it possible to measure ocean surface slope with a microwave radar at nadir probing? Int J Rem Sens 23:3251–3262
14. Karaev VYu, Kanevsky MB, Meshkov EM, Titov VI, Balandina GN (2008) Measurement of the variance of water surface slopes by a radar: verification of algorithms. Radiophys Quant Electron 51(5):1–12
15. Kaufman YJ, Smirnov A, Holben BN, Dubovik O (2001) Baseline maritime aerosol: methodology to derive the optical thickness and scattering properties. Geophys Res Lett 28:3251–3254
16. Kay S, Hedley J, Lavender S (2009) Sun glint correction of high and low spatial resolution images of aquatic scenes: a review of methods for visible and near-infrared wavelengths. Rem Sens 1(4):697–730. https://doi.org/10.3390/rs1040697
17. Khristoforov GN, Zapevalov AS, Smolov VE (1994) The relationship between the amplitudinal characteristics of the high-frequency spectral components of wind-generated waves and the wind velocity over the sea. Phys Oceanogr 5(3):221–230
18. Khristoforov GN, Zapevalov AS, Smolov VE, Feldman YuR (1994) Lidar sensing of topographic inhomogeneities on the rough sea surface. Phys Oceanogr 5(6):461–468
19. Khristophorov GN, Zapevalov AS, Babiy MV (1992) Statistics of sea-surface slope for different wind speeds. Okeanologiya 32(3):452–459
20. Kuznetsov C, Sapryikina Ya (2012) Eksperimentalnyie issledovaniya vozniknoveniya voln-ubiyts pri evolyutsii uzkogo spektra krutyih voln (Experimental study of the occurrence of freak waves during the evolution of a narrow spectrum of steep waves) Fundamentalnaya i prikladnaya gidrofizika 5(1):52–63
21. Lebedev NE, Pustovoitenko VV, Pokazeev KV, Melnikova ON (2014) Simulation of the sea surface bidirectional reflectance distribution function. Sovremennye problemy distantsionnogo zondirovaniya Zemli iz kosmosa. 11(3):310–320
22. Longuet-Higgins MS (1957) The statistical analysis of a random, moving surface. Philos Trans Royal Soc Math Phys Eng Sci 249(966):321–387. https://doi.org/10.1098/rsta.1957.0002
23. Longuet-Higgins MS (1982) On the skewness of sea-surface slopes. J Phys Oceanogr 12:1283–1291. doi: 10.1175/1520-0485(1982)012
24. Nakajima T, Tanaka M (1983) Effect of wind-generated waves on the transfer of solar radiation in the atmosphere—ocean system. J Quant Spect Rad Trans 29:521–537
25. Pokazeev KV, Zapevalov AS, Lebedev KE (2017) Measurements of sea surface slopes by laser sensing from a space vehicle. Moscow Univ Phys Bull 72(4):410–414
26. Pokazeev KV, Zapevalov AS, Pustovoytenko VV (2013) The simulation of a radar altimeter return waveform. Moscow Univ Phys Bull 68(5):420–425. https://doi.org/10.3103/S00271349 13050135
27. Pustovoytenko VV, Lebedev NE (2015) Comparison of sea surface slope statistical moments obtained by means of optical scanners and laser inclinometers. Sovremennye Problemy Distantsionnogo Zondirovaniya Zemli iz Kosmosa 12(1):102–109

28. Stewart RH (2008) Introduction to physical oceanography. Texas A & M University, Prentice Hall. p 345. doi: 10.1119/1.18716
29. Strizhkin II (2010) Analysis of optical method of determining wave slopes from photographs of glitter zone Izvestiya. Atmos Oceanic Phys 46(3):379–387
30. Tatarskii VI (2003) Multi-Gaussian representation of the Cox–Munk distribution for slopes of wind-driven waves. J Atmos Oceanic Technol 20:1697-1705
31. Torrance KE, Sparrow EM (1967) Theory for off-specular reflection from roughened surfaces. J Opt Soc Am 57(9):1105–1114
32. Viollier M, Tanré D, Deschamps P-Y (1980) An algorithm for remote sensing of water color from space. Boundary Layer Meteor 18:247–267
33. Zapevalov AS (1995) Estimation of the angular energy distribution function of the dominant sea waves. Izvestiya Akademii Nauk Fizika Atmosfery i Okeana. 31(6):835–841
34. Zapevalov AS (2000) Variability of lidar signal features when remotely sensing the sea surface. Oceanology 40(5):742–746
35. Zapevalov AS (2018) Determination of the statistical moments of sea-surface slopes by optical scanners. Atmos Oceanic Opt 31(1):91–95. https://doi.org/10.1134/S1024856018010141
36. Zapevalov AS (2020) Distribution of variance of sea surface slopes by spatial wave range. Sovremennye Problemy Distantsionnogo Zondirovaniya Zemli iz Kosmosa 17(1):211–219
37. Zapevalov AS, Lebedev NE (2014) Simulation of statistical characteristics of sea surface during remote optical sensing. Atmos Oceanic Opt 27(6):487–492
38. Zapevalov AS, Lebedev NE (2017) Impact of the sounding pulse duration of a spaceborne lidar on the shape of the pulse reflected by the sea surface. Sovremennye Problemy Distantsionnogo Zondirovaniya Zemli iz Kosmosa 14(1):80–87. https://doi.org/10.21046/2070-7401-2017-14-1-80-87
39. Zapevalov AS, Pustovoitenko VV (2010) Modeling of the probability distribution function of sea surface slopes in problems of radio wave scattering. Radiophys Quant Electron 53(2):100–110. https://doi.org/10.1007/s11141-010-9206-z
40. Zapevalov AS (2002) Statistical characteristics of the moduli of slopes of the sea surface. Phys Oceanogr 12(1):24–31
41. Zapevalov AS (2008) Statistical models of the sea surface in problems of acoustic and electromagnetic radiation scattering. Manuscript to claim the academic degree of doctor of physico-mathematical sciences on the speciality 04.00.22—geophysics. Marine Hydrophysical Institute of the National Academy of Sciences of Ukraine, Sebastopol
42. Zapevalov AS, Bol'shakov AN, Smolov VE (2011) Simulating of the probability density of sea surface elevations using the Gram–Charlier series. Oceanology 51(3):406–413
43. Zapevalov AS, Pokazeev KV (2004) Sea surface slopes statistics and its application to the laser sounding. Vestnik Moskovskogo Universita. Ser. 3 Fizika Astronomiya. 5:53–56
44. Zapevalov AS, Ratner YuB (2003) Analytic model of the probability density of slopes of the sea surface. Phys Oceanogr 13(1):1–13
45. Zhang H, Wang M (2010) Evaluation of sun glint models using MODIS measurements. J Quant Spectrosc Radiat Transfer 111(3):492–506

Chapter 8
The Distribution of the Variance of the Sea Surface Slopes on the Spatial Scales Creating Their Waves

8.1 Introduction

The solution of a number of oceanographic problems requires information on what contribution to the variance of sea surface slopes is made by waves of different scales. The nature of the scattering of radio waves is determined by the ratio between the length of the probing radio wave and the surface wavelength. In quasi-vertical sounding, sea-surface slopes created by waves whose length is comparable to and less than the length of the probe radio wave lead to diffuse scattering [2]. If the dominant mechanism is the resonance scattering mechanism, then the presence of long (compared with resonance) waves leads to a change in the local angles of incidence, since the resonance waves propagate along a curved surface [21, 24]. Slopes created by long waves must also be taken into account when analyzing microwave radiometric measurements [8].

There are several approaches to determining the variance of slopes created by long waves, i.e. waves from the longest to waves having a certain length [25]. The length is determined by the characteristics of the measuring equipment, with which the variance of the slopes is determined. In the works [4, 14] the variance of long-wavelength slopes is determined by radio sounding from a spacecraft at low angles of incidence. In the works [8, 10, 23] the dependences of the slope variance on the length of the surface waves creating them were obtained by fitting theoretical calculations to microwave radiometric measurements.

Another approach is based on the analysis of in situ measurements of sea surface slopes with various types of sensors. Laser slopemeters [11, 15], string gauge [13, 27], and wave buoys [19] are used to measure sea slopes. Depending on the design features of the measuring equipment, the slope variance obtained using it correspond to different values Λ_0.

A. Zapevalov et al., *Simulation of the Sea Surface for Remote Sensing*, Springer Oceanography,
https://doi.org/10.1007/978-3-030-58752-9_8

The aim of the research, the results of which are presented in Chapter 8, is to build the dependence of the variance of the slopes of the sea surface on the upper boundary of the range of waves that create them.

8.2 Variances of Slopes of the Sea Surface

We determine the spectral characteristics of a random wave field. Let the elevation of the sea surface $\eta(\vec{x}, t)$ be a stationary random function of the spatial coordinate $\vec{x}$ and time t. Its space-time correlation function is

$$\tilde{B}(\vec{r}, \tau) = \frac{B(\vec{r}, \tau)}{\overline{\eta^2}} \frac{\overline{\eta(\vec{x}, t)\eta(\vec{x} + \vec{r}, t + \tau)}}{\overline{\eta^2}} \tag{8.1}$$

where the bar above means averaging, $\overline{\eta^2}$ is a variance of elevations of the surface, a shift in space, t is a shift in time. The Fourier transform of a function $R(\vec{r}, \tau)$ is a three-dimensional spectrum of waves [20]

$$F\left(\vec{K}, \omega\right) = (2\pi)^{-3} \int_{-\infty}^{\infty} \int_{-\infty}^{\infty} B(\vec{r}, \tau) \exp\left(-i\left(\vec{K}\vec{r} - \omega\tau\right)\right) d\vec{r}\, d\tau \tag{8.2}$$

where $\vec{K}$ is a wave vector, ω is a cyclic frequency. Integrating the spectrum $F\left(\vec{K}, \omega\right)$ over frequencies, we obtain a two-dimensional spectrum of wave vectors

$$\tilde{\Psi}\left(\vec{K}\right) = \int_{\omega} F\left(\vec{K}, \omega\right) d\omega \tag{8.3}$$

We turn to the polar coordinate system; for this, we represent the orthogonal components of the wave vector as

$$K_x = K \cos\varphi,\ K_y = K \sin\varphi, \tag{8.4}$$

where $K = \left|\vec{K}\right|$ is wave number. We get

$$\Psi(K, \varphi) = \Psi(K_x, K_y) \frac{\partial\left(K_x, K_y\right)}{\partial(K, \varphi)} \tag{8.5}$$

where $\frac{\partial(K_x, K_y)}{\partial(K, \varphi)}$ is Jacobian.

Slopes of the sea surface are the first spatial derivatives of elevations of the sea surface. It is customary to distinguish two components of the slopes, oriented along and across the direction of the wind, which we denote by the indices "u" and "c". The spectra of slopes of the sea surface are described by the expressions [22]

$$\Upsilon_u(K) = \int_{-\pi}^{\pi} K\,\Psi(K,\varphi)\,K^2\cos^2\varphi d\varphi \tag{8.6}$$

$$\Upsilon_c(K) = \int_{-\pi}^{\pi} K\,\Psi(K,\varphi)\,K^2\sin^2\varphi d\varphi \tag{8.7}$$

where $\Psi(K,\varphi)$ is spectrum of wave numbers and directions of elevations of the surface; φ is azimuthal angle measured from the direction of the wind. Total variance

$$D(K_0) = d_u^2(K_0) + d_c^2(K_0) = \int_0^{K_0}\int_{-\pi}^{\pi} K\,\Psi(k,\varphi)K^2 d\varphi dK \tag{8.8}$$

where d_u^2 and d_c^2 are variance of the longitudinal and transverse components of the slopes. The upper limit of integration K_0 determines the boundary of the wavelength range, which corresponds to the variance of the components of the slopes.

8.3 Radar Determination of Slope Variance

The determination of the variance of the slopes based on the quasi-vertical radio sounding of the sea surface is based on the Kirchhoff method. At small angles of incidence θ, the signal reflected from the sea surface is formed by surface waves that satisfy the conditions [2]

$$\Lambda >> \lambda \tag{8.9}$$

$$k\,R\,\cos^3\theta >> 1 \tag{8.10}$$

where Λ and λ are surface and radio wave lengths, respectively; $k = 2\pi/\lambda$ is wave number of the radio wave; R is radius of curvature; θ is angle of incidence. Next, we will call long waves those waves that satisfy the condition (8.9).

In this case, the normalized backscattering cross section is described by the expression

$$\sigma = \pi\ \sec^4\theta|F_0|^2 P\left(\xi_\uparrow, \xi_\perp\right)\big|_{\xi_\uparrow = tg\theta, \xi_\perp = 0} \tag{8.11}$$

where F_0 is Fresnel coefficient; P is two-dimensional probability density function of slope; $\xi_{\uparrow}$ is slope of the sea surface in the sounding plane; $\xi_{\perp}$ is the slope of the sea surface in the orthogonal plane. The dependence of the radar signal on the probability density function of slopes of the sea surface allows us to solve the inverse problem of restoring the variance of slopes by active sensing [14].

The real sea surface is non-Gaussian. Gram-Charlier fourth-order expansion is used to express the non-Gaussian slope probability density function

$$P_{G-C}(\tilde{\xi}_c, \tilde{\xi}_u) = \frac{1}{2\pi}\exp\left[-\frac{1}{2}\left(\tilde{\xi}_c^2 + \tilde{\xi}_u^2\right)\right]\left\{1 - \frac{1}{2}\lambda_{21}H_2\left(\tilde{\xi}_c\right)H_1\left(\tilde{\xi}_u\right) - \frac{1}{6}\lambda_{03}H_3\left(\tilde{\xi}_u\right)\right.$$
$$\left.+\frac{1}{24}\lambda_{40}H_4\left(\tilde{\xi}_c\right) + \frac{1}{24}\lambda_{04}H_4\left(\tilde{\xi}_u\right) + \frac{1}{4}\lambda_{22}H_2\left(\tilde{\xi}_c\right)H_2\left(\tilde{\xi}_u\right)\right\}, \tag{8.12}$$

where $\tilde{\xi}_c = \xi_c/\sqrt{d_c^2}$, $\tilde{\xi}_u = \xi_u/\sqrt{d_u^2}$, λ_{ij} are cumulants of the crosswind and upwind components of the slopes, H_n are Hermite polynomials of order n. It is assumed that the index i refers to the crosswind component of the slopes, the index j refers to the upwind component.

One solution to the inverse problem is given in [4]. The slopes of the sea surface were determined from microwave data of precipitation radar (PR-radar) by the Tropical Rainfall Mapping Mis-sion [3]. The Ku-band PR radar mounted on the satellite is designed to measure the vertical precipitation profile. Its peculiarity is that it operates at low incidence angles and is potentially capable of providing information on the characteristics of the sea surface [6].

Dependences of slope variances on wind speed were obtained in [4]

$$D_C = 0.0189 + 0.00132\ U \tag{8.13}$$

The sounding was carried out at a frequency of 13.8 GHz ($\lambda = 2.17$ cm).

The condition is much larger in expression (8.9) is not strictly defined [8]. We will assume that expression (8.9) is satisfied if $\Lambda > n\lambda$, where n is in the range from 3 to 5. Thus, the estimates correspond to values D_C lying in the range Λ_0 from 6.5 to 10.9 cm.

8.4 Radiometric Determination of Slope Variance

With passive location (radiometry) of the sea surface, as with active, it is necessary to take into account the change in roughness created by waves of certain scales [10, 23]. The physical mechanism for generating a radiometric signal is similar to the formation of an optical signal. The difference is that when light is reflected, the conditions of specular reflection are always satisfied, and when radio sounding of the sea surface there are always waves whose length is comparable to or less than the length of the radio wave. These waves, which do not satisfy condition (8.9), create diffuse scattering.

In radiometric measurements, to determine the variance of the slopes created by long waves, a semi-empirical function is used, constructed by minimizing the discrepancies between theoretical calculations and data from radiometric measurements. It is based on the assumption that the variance of the slopes that determine the radiometric signal can be described by a universal function of the frequency (or length) of the radio wave $\chi(f)$ as part of the total variance $D(K_0 = \infty)$

$$\chi(f) = \begin{cases} 0.3 + 0.02f & npu\ f \leq 35 \\ 1 & npu\ f > 35 \end{cases} \tag{8.14}$$

where f is frequency of a radio wave of GHz. Thus, variances of long waves, for which the upper limit of wave numbers is equal K_0, are described by the expression

$$D(K_0, U) = \chi(K_0)\, D(K_0 = \infty, U) \tag{8.15}$$

where $D(K_0 = \infty, U)$ is the dependence of the variance of the slope on the wind speed, obtained from the processing of aerial photographs in [7].

Within the framework of this approach, the values of the function $\chi(K_0)$ for three wavelengths: 6, 2.25, and 0.8 cm were calculated in [8]. Experimental studies were carried out on an oceanographic platform located on the Black Sea. The procedure included fitting the values of the variances of the slopes in order to achieve maximum fit in the calculation model for both vertical and horizontal polarizations. Further, as in the interpretation of active sensing data, we assume that the radiometer signal is formed by surface waves whose length is more than 3–5 times the length of the recorded radio wave.

8.5 In Situ Slope Measurements

To measure the slope of the sea surface under field conditions, slopemeters are used, the principle of operation of which is based on various physical effects. A consequence of the difference in the sensors used was a significant difference in the results obtained by different authors. The most effective for measuring slopes are optical methods that use laser sensors [15]. The principle of operation of the laser slopemeter is based on measuring in two planes the angles of deviation of the laser beam from the vertical when it passes the water-air interface. The diameter of the laser spot on the surface is about 0.2 cm. Thus, it can be assumed that the contribution to the variance is made by waves whose length is more than 0.4–0.6 cm.

Let us consider the results of studies of the structure of the sea surface obtained using laser slopemeter using two different methods. In the first case, measurements were carried out from a moving vessel, in the bow of which a laser slopemeter was

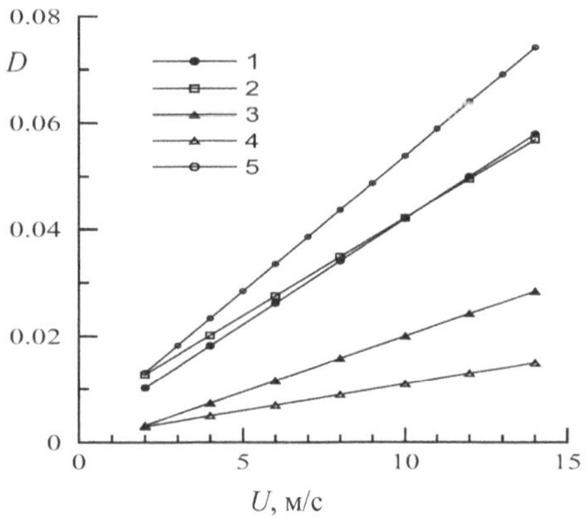

Fig. 8.1 Dependences of the variance of sea surface slopes, determined from in situ measurements: line 1 and 2 are measurements by a moving and stationary laser slopemeter [11, 15]; 3 is measurements by string gauge [13]; line 4 is measurements by a wave buoy [19]; line 5 is aerial photography of the sun glint [7]

installed [11]. The experiment was conducted on the estuary of a river that flows into the Georgia Strait. The dependence on the wind speed of the variance of the slopes is described by the expression

$$D_{L1} = 0.0023 + 0.00397\ U \tag{8.16}$$

In the second case, measurements were carried out from a stationary base, which is the oceanographic platform [15]. The experiments were conducted on the Black Sea

$$D_{L2} = 0.00546 + 0.00367\ U \tag{8.17}$$

As can be seen from Fig. 8.1, the dependences of the variance of the slopes on the wind speed, obtained from measurements at a point and from a moving carrier, were close to each other

Slope variance measurements made using differential string gauge (Kalinin and Leikin [13] lead to regression dependence

$$D_A = -0.001 + 0.0021\ U \tag{8.18}$$

The variance values D_A are significantly lower than the values obtained using laser slopemeters. Sensors in differential string slopemeters are vertically stretched strings (strings 0.3 mm in diameter were used). Slopes are defined as the ratio of the difference in surface levels at two spatially separated points to the distance between these points. In this experiment, the distance between the string sensors was 20 cm or 40 cm. The experiments were conducted on the Caspian Sea.

The lowest values of the variance of the slopes give measurements using floating on the surface of buoys of special designs [19].

$$D_B = 0.001 + 0.001\ U \tag{8.19}$$

The sensor in this type of meter is the buoy body. The frequency range of the measured slopes is limited from above at a frequency of about 0.5 Hz, which for surface waves in deep water corresponds to a wavelength of about 6 m. As in the above cases of in situ measurements, it is necessary to take into account the uncertainty in estimating the length of short surface waves affecting the variance of slopes.

8.6 Comparison of Variances of Slopes of the Sea Surface

The lengths of electromagnetic waves of the optical range are much shorter than the lengths of the waves present on the sea surface; therefore, all waves present on the sea surface form an optical signal. Optical measurements can be accepted $K_0 = \infty$. The most often used regression dependence of the variance of slopes on wind speed, obtained on the basis of aerial photographs of the sea surface [7]

$$D(K_0 = \infty, U) = D_{CM} = 0.003 + 0.00508\ U \tag{8.20}$$

where U is wind speed. Let us consider the ratio of the variance of slopes created by long waves and the total variance created by the waves of all scales present on the sea surface $\chi = D/D_{CM}$.

The dependences of the variances of the slopes of the sea surface on the wind speed, calculated from the data of remote sensing and from in situ measurements, are usually presented in the form of a linear regression

$$D(U) = a + bU \tag{8.21}$$

The regressions (8.13), (8.16)–(8.20) used in this work are presented in the same form. Comparing the dependences of the slope variances on the wind speed obtained by different methods, we can see that in the range of weak winds the parameter χ depends on the wind speed. This is due to the fact that linear regression incorrectly describes changes in the variance of slopes in a wide range of wind speeds. The increase in the variance of slopes with increasing wind in the region of weak winds occurs much faster than with a strong wind [26]. As a result, when approximating a nonlinear dependence of a linear function, a constant term a appears. At a wind speed U_0 above 7–8 m/s, the first term of linear regression can be neglected, while the ratio of the variance of the slopes determined using different types of equipment turns out to be independent of the wind speed.

A summary graph of the dependence of the parameter χ_i on the surface wavelength Λ_0 for wind speeds $U > U_0$ is shown in Fig. 8.2. The type of measuring equipment and the source of information on the basis of which the values χ_i were calculated are listed in Table 8.1. When constructing Fig. 8.2 it was taken into account that the concept is much larger in (8.9) is not strictly defined. Therefore, dependences (8.14) correspond to two curves 1, calculated for situations when the signal is formed by surface waves 3 or 5 times the length of the probing radio waves. This circumstance

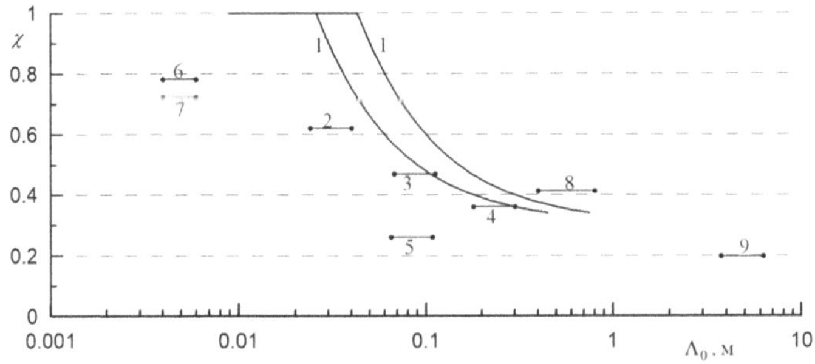

Fig. 8.2 Empirical estimates of the fraction of the total variance of the slopes of the sea surface, determined in the range of wavelengths from the main energy-carrying waves to waves, by length Λ_0. The type of measuring equipment and the source of information are shown in Table 8.1

Table 8.1 Types of measuring equipment and information sources used in the construction of Fig. 8.2

#	Measuring equipment	References
1	Radiometer	[23]
2	Radiometer, $\lambda = 0.8$ cm	[8]
3	Radiometer, $\lambda = 2.25$ cm	
4	Radiometer, $\lambda = 6$ cm	
5	Radar	[4]
6	Laser slopemeter	[11]
7	Laser slopemeter	[15]
8	String gauge	[13]
9	Wave buoy	[19]

was taken into account when constructing parameter estimates χ_i ($i = 2, 3, 4$, and 5) based on remote sensing data. The surface wavelength Λ_0 is also not strictly defined in situ measurements. For this measurement data on the graph, the points are replaced by segments, the boundaries of which are selected according to the above description of the measuring equipment. As can be seen from Fig. 8.2, parameter χ_i estimates are in fairly good agreement with each other. The parameter χ_5 is slightly deviated, the values of which seem to be underestimated.

Based on the data presented in Fig. 8.2, we construct a regression that describes the dependence $\chi = \chi(\Lambda_0)$. The upper limit of the range in which waves can exist on the surface of a liquid is due to viscous dissipation. This limit $K_{dis} \approx 6283$ is rad/m [1], which corresponds to a wavelength of $\Lambda_{dis} \approx 0.001$ m. Since the length of electromagnetic waves in the optical range is much shorter Λ_{dis}, we can assume that $D(K_{dis}) = D_{CM}$. Thus, we get the point $\chi(\Lambda_0 = \Lambda_{dis}) = 1$, which is also used in the construction of the regression $\chi = \chi(\Lambda_0)$.

The analysis presented in Fig. 8.2 estimates of the share of the total variance of long wavelength slopes from the measurement data showed that the dependence $\chi = \chi(\Lambda_0)$ can be approximated by regression in the form of a power function.

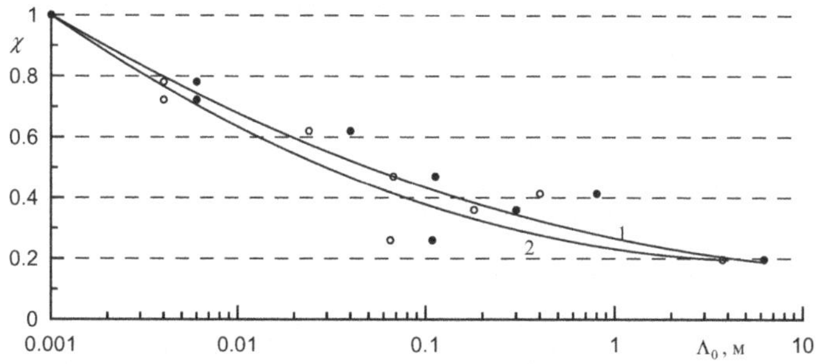

Fig. 8.3 Regression dependences of the fraction of the total variance of the slopes of the sea surface χ, defined in the wavelength range from the main energy-carrying waves to waves, with the length Λ_0: curve 1 is regression (8.23); curve 2 is regression (8.22); symbols ○ and ● show the minimum and maximum values Λ_0

Since in all cases the measurements, the results of which are presented in Fig. 8.2, the length Λ_0 is not uniquely determined, two dependencies corresponding to the minimum and maximum values Λ_0 were calculated

$$\chi(\Lambda_0) = 0.260\Lambda_0^{-0.194} \pm 0.082 \tag{8.22}$$

$$\chi(\Lambda_0) = 0.294\Lambda_0^{-0.177} \pm 0.082 \tag{8.23}$$

These regressions are shown in Fig. 8.3. If we exclude from the calculations the estimate χ_5 obtained by the results of [4], then

$$\chi(\Lambda_0) = 0.285\Lambda_0^{-0.182} \pm 0.045 \tag{8.24}$$

$$\chi(\Lambda_0) = 0.320\Lambda_0^{-0.166} \pm 0.045 \tag{8.25}$$

Calculation of the dependence $\chi = \chi(\Lambda_0)$ without a parameter χ_5 reduces the mean-square spread by almost half.

When comparing Figs. 8.2 and 8.3 it is seen that the expression (8.14) proposed on the basis of the data of radiometric measurements significantly overestimates the values χ in the region of $\Lambda_0 < 0.1$ m. From Fig. 8.3 it also follows that the contribution to the variance of wave slopes longer than 10 m is less than 20%.

Another factor to consider is the following. Any spectral model of the surface wave field is built for certain conditions and within the framework of certain assumptions that are not always possible to control. This circumstance limits the possibilities of their use for the analysis and interpretation of remote sensing data. In this work, to construct the dependence $\chi = \chi(\Lambda_0)$, we used data obtained by remote sensing and in situ measurements. Unfortunately, in these works there is no information on how the variance of the slopes varies at different stages of the development of the wave field. Equations (8.22) and (8.23) correspond to an ensemble of situations in which regression equations $D = D(U)$ were obtained in different experiments.

8.7 Calculation of the Distribution of Surface Variances Based on Spectral Model

In recent years, spectral models defined in the space of wave numbers have been widely used to analyze the interaction of radio waves with the sea surface [4, 5]. When constructing mathematical models, the lack of information about the fine topographic structure of the sea surface was compensated by laboratory data [12, 16]. As a rule, they are built on the basis of the kinetic equation, their parameters are adapted to the data of laboratory experiments. It is assumed that the variance of slopes calculated over the entire range of wave numbers should correspond to the variance determined by remote sensing in the optical range [7].

The spectral energy of sea waves decreases rapidly with decreasing wavelength, therefore, for its analysis, it is convenient to use the curvature spectra $Cr(k)$ associated with the elevation spectrum by the relation

$$Cr(K) = K^4 S(K) \tag{8.26}$$

where $S(K) = \int_{-\pi}^{\pi} \Psi(K, \varphi) d\varphi$. Curvature is the second spatial derivative of elevations of the sea surface.

Models of curvature spectra constructed to analyze the interaction of radio waves with the sea surface indicate the existence of a secondary peak in the gravitational-capillary range of surface waves (see [5, 17]), moreover, values $Cr(K)$ in its vicinity depend on wind speed. The secondary peak grows rapidly at low wind speeds, at high it approaches the saturation state that is observed at wind speeds above 6–8 m/s [1, 9]. Saturation leads to the fact that the shape of the spectrum $Cr(K)$ in the gravitational-capillary range varies slightly with increasing wind speed.

In this work, we will use the one-dimensional model of the broadband spectrum of surface waves proposed in [18]. Like most similar models, it includes two components

$$S(K) = S_1(K) + S_2(K) \tag{8.27}$$

The first component $S_1(K)$ is based on field measurements. It describes the spectrum on scales from the scale of dominant surface waves to the lower boundary of the range of short gravitational waves

$$S_1(K) = S_P(K) \frac{1}{\left(1 + (K/K_0)^2\right)} \tag{8.28}$$

where $S_P(K) = \tilde{a}_p K^{-4}$ is Phillips spectrum in the equilibrium range [20]; $\left(1 + \left(K/K_0\right)^2\right)^{-1}$ is the low-pass filter constructed in [1] ($K_0 = 100$ rad/m) is the upper limit of the wave number range for gravitational waves). The lower limit of the range in which the spectrum $S_1(K)$ is modeled is determined from the condition

$$K_p = g/(1.2\,U)^2 \tag{8.29}$$

Here, the parameter K_p corresponds to the scale of the spectral peak.

The second component $S_2(k)$ describes the spectrum in the range of gravitational-capillary waves

$$S_2(K) = a_{L1}\,K^{-4}\left(\frac{u_* - u_{0*}}{C}\right)\left[1 - \exp\left(-\frac{C^2}{a_{L2}\,W_{10}^2}\right)\right]$$
$$\exp\left[-a_{L3}K^{2.5}(u_* - u_{0*})^{-0.75}\right] \tag{8.30}$$

where $a_{L1} = 1/320$, $a_{L2} = 0.0002$ and $a_{L3} = 0.0011(\text{cm/rad})^{2.5}(\text{cm/s})^{0.75}$ are constants; u_* is friction velocity; $u_{0*} = 5$ cm/s is the threshold value of the friction velocity; C is the phase velocity. The relationship of wind speed and friction speed is set using the drag coefficient in the form

$$C_d = 10^{-3}\left(2.717\,U^{-1} + 0.142 + 0.0761\,U\right) \tag{8.31}$$

where U is wind speed at a height of 10 m.

The one-dimensional curvature spectra $Cr(k)$ calculated using the models (8.27)–(8.31) are shown in Fig. 8.4.

To estimate the contribution of waves of different scales to the variance of the slopes, we introduce the integral parameter [28]

$$\chi_M(K) = \int_{K_p}^{K_0} Ss(K)dk \Bigg/ \int_{K_p}^{K_{dis}} Ss(K)dk \tag{8.32}$$

where $Ss(K)$—range of slopes. From the spectral model proposed in [18], it follows that the parameter values χ_M change with changing wind speed. The nature of these changes is shown in Fig. 8.5, which shows the dependencies $\chi_M = \chi(\Lambda_0)$ for different wind speeds.

Hereinafter, the index M shows that this parameter was obtained as a result of calculations according to the spectral model [18], the index E shows that the parameter χ was obtained from the measurement data. It should be remembered that the dependences $\chi_E = \chi_E(\Lambda_0)$ obtained in this work are valid for wind speeds above 7–8 m/s, i.e. precisely in the range where, according to spectral models, a saturation state takes place [5, 9, 17]. To build the dependence $\chi_E = \chi_E(\Lambda_0)$ on weaker winds, additional research is needed.

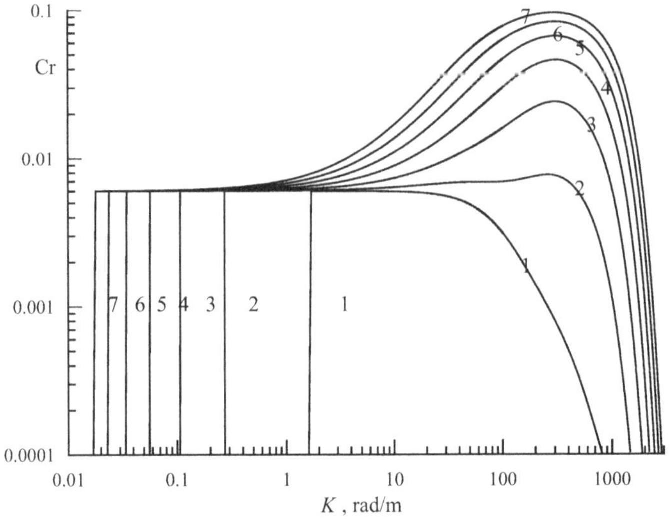

Fig. 8.4 Model spectra of curvature of the sea surface $Cr(K)$. Curves 1–7 correspond to wind speeds from 2 to 20 m/s in increments of 3 m/s

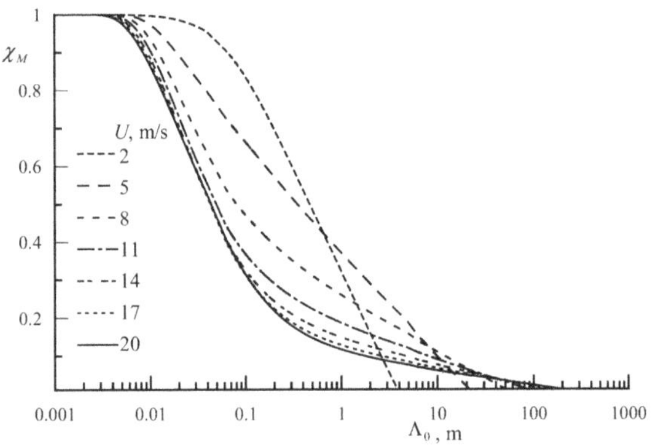

Fig. 8.5 Dependencies $\chi_M = \chi_M(\Lambda_0)$

Let us compare the model estimates of the parameter χ_M with the dependences of this parameter χ_E on Λ_0 obtained on the basis of measurements [28]. As already noted in determining the variance of slopes from remote sensing data and from in situ measurements, there is always some uncertainty in determining the boundary value of the length Λ_0. Therefore, each parameter χ_E in Fig. 8.6 are assigned two values Λ_0 corresponding to the maximum and minimum values Λ_0.

Shown in Fig. 8.6 parameter χ_E estimates are calculated on the basis of data obtained by apparatus of various types. The numbers around the ratings χ_E correspond to the numbers in the Table 8.1, where the works are indicated whose results were used in the calculations. Although in general there is a correspondence between

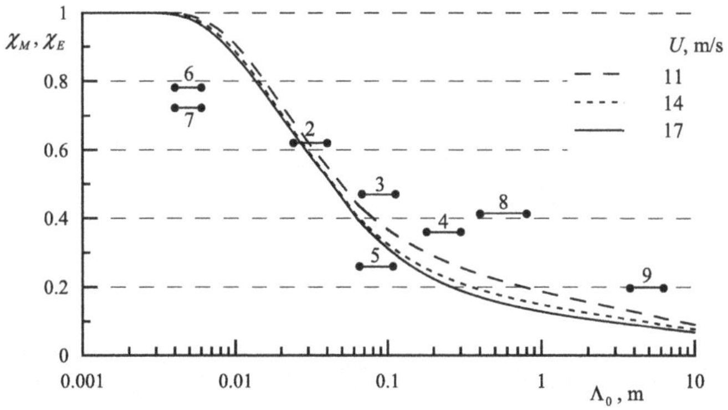

Fig. 8.6 Dependences of the parameters χ_M and χ_E on the wavelength Λ_0. The numbers near the estimates χ_E correspond to the Table 8.1

the theoretical and experimental values of the parameter χ_E, there is a significant difference at wavelengths shorter than 0.01 m.

Note that the spectra of slopes of surface waves, as well as the spectra of elevations, depend not only on wind speed, but also on the stage of wave development [8]. This effect is most noticeable on the scales of dominant waves, with a decrease in the wavelength it weakens. The shorter the wavelength, the shorter the relaxation time, however, changes in the wave spectra at a fixed wind speed at different stages of development should also occur in the short-wavelength region. Unfortunately, the data available to us do not yet allow us to estimate this effect in the wavelength range in which the dependence $\chi_E = \chi_E(\Lambda_0)$ is analyzed here.

Shown in Fig. 8.6 dependences $\chi_M = \chi_M(\Lambda_0)$ show that the most dramatic changes χ_M occur in the range 0.01–0.1 m. In [25], an attempt was made to simulate this effect by setting regressions in a form that included a larger number of parameters

$$\chi_E(\Lambda_0) = 0.231 - 0.0923\log(\Lambda_0) + 0.0546\ \log^2(\Lambda_0) \tag{8.33}$$

$$\chi_E(\Lambda_0) = 0.266 - 0.129\log(\Lambda_0) + 0.0386\ \log^2(\Lambda_0) \tag{8.34}$$

The standard deviations for both regressions were ±0.08, i.e. practically coincides with such a characteristic for regressions (8.22) and (8.23). The effect of significant changes in the parameter χ_E in the range of 0.01–0.1 m, as shown in Fig. 8.7 is not observed. Thus, to calculate the variance of slopes in applications related to diffraction of waves on the sea surface, it is advisable to use regressions that have a simple analytical form (8.22) and (8.23).

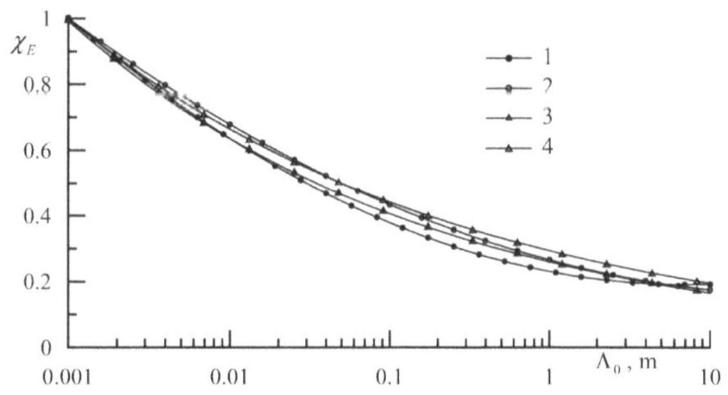

Fig. 8.7 Regression $\chi_E = \chi_E(\Lambda_0)$. Curves 1-4 correspond to Eqs. (8.33), (8.34), (8.22) and (8.23)

8.8 Conclusion

The interaction of radio waves with the sea surface is described in the framework of a two-scale model. The boundary dividing the two ranges of irregularities depends on the length of the probe radio wave and on the angle of incidence. The existence of several physical mechanisms that determine the diffraction of radio waves on a rough surface makes the problem of estimating the variance of the slopes of the sea surface in different wave ranges relevant.

The solution to this problem is complicated by the lack of sufficient information about the real variability of the slopes of the sea surface, which is largely due to the technical complexity of their measurements in the field. In Chapter 8, to solve it, the results of measurements carried out with various types of equipment were used. The data of remote sensing of the sea surface (radar, radiometric, and optical measurements), as well as in situ measurements (measurements by laser slopemeters, string gauge, wave buoys) are analyzed. The results of experimental studies were compared with the results of numerical modeling, based on the spectral model of elevations of the sea surface, built in the space of wave numbers.

An integrated approach made it possible to construct functions $\chi(\Lambda_0)$ describing the variance of slopes of the sea surface created by waves in the range from the main energy-bearing waves to a given length Λ_0, as part of the total variance of slopes created by all waves present on the sea surface. It is shown that the dependence $\chi = \chi(\Lambda_0)$ can be described by a power function. It is also shown that the contribution to the variance of slopes, which are produced by waves longer than 10 m, is less than 20%.

References

1. Apel JR (1994) An improved model of the ocean surface wave vector spectrum and its effects on radar backscatter, J Geophys Res 99(C8):16269–16291
2. Bass FG, Fuks IM (1972) Rasseyaniye voln na statisticheski nerovnoy poverkhnosti (Wave scattering from statisti-cally rough surface). Nauka, Moscow, p 424

3. Cecil DJ, Goodman SJ, Boccippio DJ, Zipser EJ, Nesbitt SW (2005) Three years of TRMM precipitation features. Part I: Radar, Radiometric, and lightning characteristics. Month Mfeather Rev 133(3):543–566
4. Chen P, Yin Q, Huang P (2015) Effect of non-Gaussian properties of the sea surface on the low-incidence radar backscatter and its inversion in terms of wave spectra by an ocean wave. Chinese J Oceanol Limnology 33(5):1142–1156
5. Cheng Y, Liu Y, Xu Q (2006) A new wind-wave spectrum model for deep water. Indian J Mar Sci 35(3):181–194
6. Chu X, He Y, Chen G (2012) Asymmetry and anisotropy of microwave backscatter at low incidence angles. IEEE Trans Geosci Remote Sensing 50:4014–4024
7. Cox C, Munk W (1954) Measurements of the roughness of the sea surface from photographs of the sun glitter. J Opt Soc America 44(11):838–850
8. Danilychev MV, Nikolaev AN, Kutuza BG (2009) Application of the kirchhoff method for practical calculations in microwave radiometry of wavy sea surface. J Commun Technol Electron 54(8):869–878
9. Elfouhaily T, Chapron B, Katsaros K, Vandemark D (1997) A unified directional spectrum for long and short wind-driven waves. J Geophys Res 102(C7):15781–15796
10. Hollinger JP (1971) Passive microwave measurements of sea surface roughness. IEEE Trans Geosci Electr GE-9(3):165–169
11. Hughes BA, Grant HL, Chappell RWA (1977) A fast response surface-wave slope meter and measured wind-wave components. Deep-Sea Res 24(12):1211–1223
12. Jähne B, Riemer KS (1990) Two-dimensional wave number spectra of small-scale water surface waves. Geophys Res 95:11531–11546
13. Kalinin SA, Leikin I (1988) Izmereniye uklonov vetrovykh voln v Kaspiyskom more (A. Measurement of the slopes of wind waves in the Caspian Sea). Izvestiya of the academy of sciences of the USSR, Atmos Oceanic Phys 24(11):1210–1217
14. Karaev VYM. Panfilova MA, Balandina GN, Chu X (2012) Retrieval of the slope variance by microwave measurements. Issledovanie Zemli iz Kosmosa 4:62–77
15. Khristoforov GN, Zapevalov AS, Babii M (1992) Statistics of sea-surface slope for different wind speeds. Okeanologiya 32(3):452–459
16. Klinke J, Jähne B (1992) Two-dimensional wave number spectra of short wind waves. Results from wind wave facilities and extrapolation to the ocean. Proceeding of SPIE 1749, optics of the air-sea interface: theory and measurements, proceeding of SPIE 1749, international society for optics engineering, San Diego 1–13 (1992)
17. Kudryavtsev VN, Markin VK, Chapron B (1999) Coupled sea surface-atmosphere model 2, spectrum of short wind waves. J Geophys Res 104:7625–7639
18. Liu Y, Su MY., Yan XH, Liu WT (2000) The mean-square slope of ocean surface waves and its effects on radar backscatter. J Atmos Oceanic Technol 17:1092–1105
19. Longuett-Higgins MS, Cartwrighte DE, Smith ND (1963) Observation of the directional spectrum of sea waves using the motions of the floating buoy. In: Proceeding of conference ocean wave spectra, englewood cliffs. NY Prentice Hall, pp 111–132
20. Phillips OM (1980) The dynamics of the upper ocean. Cambridge Univesity Press, 344 p
21. Plant WJ (2002) A stochastic, multiscale model of microwave backscatter from the ocean. J Geophys Res 107(C9):3120. https://doi.org/10.1029/2001JC000909
22. Plant WJ, Irisov V (2017) A joint active/passive physical model of sea surface microwave signatures. J Geophys Res Oceans 122:3219–3239. https://doi.org/10.1002/2017JC012749
23. Wilheit TT (1979) A model for the microwave emissivity of the ocean's surface as a function of wind speed. IEEE Trans Geosci Electron GE-17(4)
24. Zapevalov AS (2009) Bragg scattering of centimeter electromagnetic radiation from the sea surface: the effect of waves longer than Bragg components. Izvestiya Atmos Ocean Phys 45(2):253–261
25. Zapevalov AS (2020) Distribution of variance of sea surface slopes by spatial wave range. Sovremennye problemy distantsionnogo zondirovaniya Zemli iz kosmosa 17(1):132–149. https://doi.org/10.21046/2070-7401-2020-17-1-211-219

26. Zapevalov AS (2002) Statistical characteristics of the moduli of slopes of the sea surface. Phys Oceanogr 12(1):24–31
27. Zapevalov AS, Bolshakov AN, Smolov VE (2009) Studying the sea surface slopes using an array of wave gauge sensors. Oceanology 49(1):31–38
28. Zapevalov AS, Knyazkov AS, Shumeyko IP (2020) Describtion of sea surface slopes in applications related to radio wave reflection. J Radio Electron (Zhurnal Radioelektroniki) 4. https://doi.org/10.30898/1684-1719.2020.4.8

Chapter 9
Study of the Variability in the Ocean-Atmosphere Interface as a Reflective Surface of Light

9.1 Introduction

Optical images from optical scanners mounted on spacecraft are an effective tool for ocean research [19–22, 37]. Several physical mechanisms form an optical scanner signal [14, 25, 39]. Such mechanisms are: reflection by the sea surface of direct solar radiation, reflection by the sea surface of solar radiation scattered by the atmosphere, reflection by the sea surface in any direction of direct solar radiation, further scattered by the atmosphere towards the satellite, scattering of solar radiation by the atmosphere towards the satellite, scattering from the water column of solar radiation penetrated under the ocean-atmosphere boundary. In recent decades, active means of sensing the ocean using lidars mounted on spacecraft have been actively developed [13, 24, 32].

The effective use of optical remote sensing requires detailed information about the reflection of light from the sea surface [43]. The main elements of the sea surface determining light scattering and reflection are short gravity, gravity-capillary and capillary waves [33]. Having relatively low energy, the waves of these ranges significantly change their characteristics under the influence of processes that occur in the boundary layers of the atmosphere and ocean. This property is the basis for remote measurement methods, most of which only carry information about the sea surface.

Indication of laser glint when probing the sea surface allows obtaining information about small-scale components of wind waves [44]. This method is successfully applied both in laboratory experiments and in field research carried out on board a ship, an aircraft, a satellite and a fixed base [6, 17, 24, 26]. One of the modifications of the method of laser glint indication is measurements in the scanning mode [36].

When interpreting the results of laser sensing of the sea surface, a priori models (spectra, probability density functions, etc.) are used, the adequacy of which is not usually checked for specific situations. Direct in situ measurements showed strong variability of short surface waves and their stochastic dependence on wind speed

A. Zapevalov et al., *Simulation of the Sea Surface for Remote Sensing*, Springer Oceanography,
https://doi.org/10.1007/978-3-030-58752-9_9

[42]. Slices and ripple areas simultaneously present on the sea surface in low winds differ not only in the spectral density level of short-period components but also in the distribution of probability density of slopes [9, 15].

The variety of factors forming the fine structure of the sea surface determines the stochastic nature of the relationship between its parameters and those recorded by laser probing. This creates the need for complex experiments in which remote sensing is accompanied by careful control of the sea surface parameters.

9.2 Measuring Equipment and Measurement Conditions

Integrated studies of the ocean-atmosphere interface as a reflective surface of light were carried out in the Black Sea during summer and autumn on a stationary oceanographic platform of the Marine Hydrophysical Institute [43]. The oceanographic platform is located off the southern coast of Crimea, 500 m from the coast at a depth of about 30 m [23]. This depth for the Black Sea corresponds to the "deep water" condition, i.e. the influence of the bottom can be neglected. During the measurement period, water-air stratification was either poorly stable or close to neutral. Measurements—were made both during daytime and at night.

Laser sensing of the sea surface was carried out simultaneously with measurements of its elevation and slope and was accompanied by measurements of wind speed. A broadband string gauge was used to determine surface wave parameters. The sensor is a vertically stretched string of nichrome wire with a diameter of 0.2 mm. The frequency range of the waveguide is 0.05–30 Hz. The wind speed meter was installed at a height of 18 m. All sea surface parameters and wind speed were measured simultaneously in 30 s sessions. The series of measurement sessions lasted from 1 to 7 h.

The principle of operation of the two-dimensional laser slope meter is based on measuring the deviation of the laser beam from the vertical when passing from under water of the rough water-air interface. Its design allows to measure the slopes in two orthogonal planes [15]. Calibration of the laser inclinometer is performed in laboratory conditions.

A stationary oceanographic platform, like any physical object, creates disturbances in the aquatic environment. Lidar and wave sensors were placed on long booms oriented towards the open sea. This allowed measurements to be made outside the perturbations created by the supports of the ocean platform. The distance from the measurement point to the nearest support was over 6 m.

For indication of lidar glint reflectance was used, optical scheme of which is shown in Fig. 9.1 [43]. The source and the receiver of the optical signal are combined. The source of optical signal is HeNe laser with power of 4 mW (1). The laser beam with a prism (3) is directed to the water surface through the lens (4). In case of glint reflectance, the reflected beam in the lens with the same prism (4) is directed to the photoreceiver (2). The lens has a diameter of 50 mm.

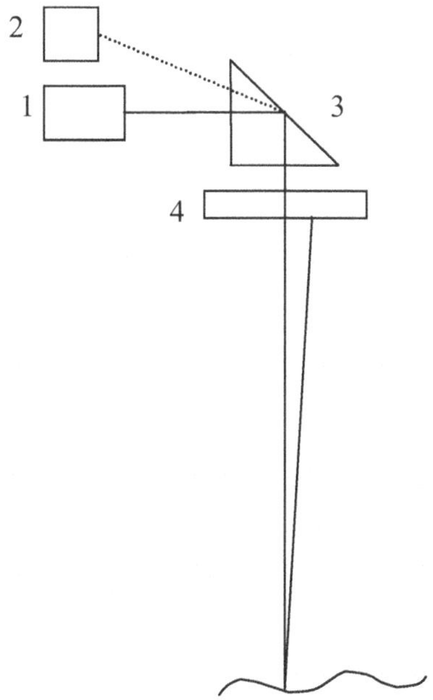

Fig. 9.1 Optical lidar circuit for laser sensing of the sea surface

When the laser beam falls vertically on the undisturbed water surface backwards, ~2% of its energy is reflected back [24]. If the surface is rough, the share of energy reflected backward decreases. In order to be able to register reflections of the mirror reflection in strong winds and significant surface disturbances, a photoreceiver was used, which had a sensitivity threshold corresponding to 0.06% of the energy of the laser beam falling on the sea surface.

Laser probing during daytime can have effects created by sun glint. In order to eliminate these effects, the laser signal was subjected to amplitude modulation with a frequency of 100 kHz. The optical signal received by lidar was converted into an electrical signal, which was then subjected to linear detection, and then only the envelope of this signal was analyzed. Thus, the information obtained can be considered as a result of measurements with a continuous laser.

In experiments on an oceanographic platform, the instrument was installed at height 5 or 12 m. The bulk of the data analysed here were obtained at an height of 5 m. At the device placement height of 5 m and vertical laser beam orientation the spot diameter on the undisturbed surface was equal to 5 mm. Further we will consider the influence of the lidar placement height on the glint statistics.

In the aperture of the photoreceiver at vertical probing only those glint that are formed by the sites, the local slope of which is less than some critical value ξ_K. From simple geometrical constructions it is possible to show that the slope value is determined by two parameters. These parameters are the aperture diameter of the photoreceiver d_F and the distance from the photoreceiver to the reflecting surface h [1].

$$\xi_K = \frac{d_F}{4 \cdot h} \tag{9.1}$$

When measuring from a height of 5 m and aperture diameter of 5 cm we obtain: $\xi_K = 0.0025$.

In Cartesian coordinate system it is accepted to determine the components of slopes oriented in upwind ξ_u and crosswind direction ξ_c. When probing vertically, the azimuthal orientation of the slope of the site on which the laser beam falls does not affect the glint registration. Let's move from Cartesian coordinate system to polar coordinate system

$$\xi_c = \xi_m \sin\varphi, \quad \xi_u = \xi_m \cos\varphi \tag{9.2}$$

where $\xi_m = \sqrt{\xi_u^2 + \xi_c^2}$—slope module; φ—azimuth angle.

When continuously probing the sea surface with a device with a combined transmitter and receiver, the recorded signal represents a sequence of electric pulses corresponding to the mirror reflections of the laser beam from the sites whose slope meets the condition of

$$\xi_m < \xi_K \tag{9.3}$$

A sequence of pulses is analyzed further. The operation of the data collection and preprocessing system was structured as follows [41].

Three parameters were calculated from the data of each measurement session. These parameters are: number of pulses N, their total duration

$$\tau_\Sigma = \sum_{n=1}^{N} \tau_n \tag{9.4}$$

where τ_n is the duration of the n-th impulse; and also the integral intensity

$$I_\Sigma = \int_0^T a(t)\,dt \tag{9.5}$$

where T is duration of the measurement session; $a(t)$ is function describing the pulse shape; t is time. There is no signal in the time intervals between the recorded glint, i.e. $a(t) = 0$. Then by three registered parameters N, τ_Σ and I_Σ the average per session characteristics were calculated:

- glint rate $F = N/T$;
- average glint duration $\tau = \tau_\Sigma / N$;
- average glint intensity $I = I_\Sigma / \tau_\Sigma$;

- the probability of the reflected beam hitting the aperture of the photoreceiver τ_{Σ}/T.

These characteristics have been further analyzed.

Note the fundamental difference in laser sensing from a spacecraft and an oceanographic platform. In the first case, the diameter of the laser spot on the sea surface is dozens of meters, and this spot always contains points that meet the condition (9.3). When probing from low altitude, the laser spot diameter is equal to several millimeters, and the mirror reflection condition is fulfilled only at certain points in time.

9.3 The Possibility of Determining the Slope of the Sea Surface by Indicating the Reflective Reflections of a Mirror

Let's consider the possibility of solving the inverse problem, restoring the sea surface slope characteristics based on the glint indication. The idea of the method for determination of sea surface slopes on the basis of measuring the duration of glint reflection was proposed in [1]. The essence of it is that as the surface slopes have Gaussian or near-Gaussian distribution, then, measuring the probability of getting the slope values in some given interval, it is possible to determine the standard-square slope.

The modeling of statistical characteristics of a random moving Gaussian surface has shown that the form of probability distribution of the slope module depends on the parameter of three-dimensionality of the wave field (Longuet-Higgins 1957).

$$\gamma = \sqrt{\overline{\xi_c^2}}/\sqrt{\overline{\xi_u^2}} \tag{9.6}$$

where $\overline{\xi_u^2}$ and $\overline{\xi_c^2}$ is the upwind and crosswind variance of the slope components.

If the slope components are described by the Gaussian distribution, we obtain in the polar coordinate system

$$P(\xi_m, \alpha) = \frac{\xi_m}{2\pi\sqrt{\overline{\xi_u^2}\,\overline{\xi_c^2}}} \cdot \exp\left[-\frac{1}{2} \cdot \left(\frac{\xi_m^2 \cdot \cos^2\varphi}{\overline{\xi_u^2}} + \frac{\xi_m^2 \cdot \sin^2\varphi}{\overline{\xi_c^2}}\right)\right] \tag{9.7}$$

The expression (9.7) after integration in all directions allows determining the probability density ξ_m

$$P(\xi_m) = \int_0^{2\pi} \frac{\xi_m}{2\pi\sqrt{\overline{\xi_u^2}\,\overline{\xi_c^2}}} \cdot \exp\left[-\frac{1}{2} \cdot \left(\frac{\xi_m^2 \cdot \cos^2\alpha}{\overline{\xi_u^2}} + \frac{\xi_m^2 \cdot \sin^2\alpha}{\overline{\xi_c^2}}\right)\right] d\alpha \tag{9.8}$$

Introduce normalizing

$$\tilde{\xi}_m = \frac{\xi_m}{\sqrt{\overline{\xi_u^2} + \overline{\xi_c^2}}} \tag{9.9}$$

It follows from (9.8)

$$P\left(\tilde{\xi}_m\right) = \tilde{\xi}_m\left(\gamma + \gamma^{-1}\right)\exp\left[-\frac{\tilde{\xi}_m^2\left(\gamma + \gamma^{-1}\right)}{4}\right] I_0\left[\frac{\tilde{\xi}_m^2\left(\gamma^{-2} - \gamma^2\right)}{4}\right] \tag{9.10}$$

where I_0 is the Bessel function zero order. Thus, the only parameter that determines the type of distribution of the normalized slope module $\tilde{\xi}_m$ is the three-dimensionality index γ.

According to the statistics of mirror reflection glint the probability $\Phi(\xi_K)$, i.e. the probability of the reflected beam hitting the aperture of the photoreceiver, can be defined as [1].

$$\Phi(\xi_K) = \tau_\Sigma / T \tag{9.11}$$

Let's analyze the distribution ξ_m in two limiting cases: $\gamma = 0$ and $\gamma = 1$. At $\gamma = 0$ (two-dimensional wave field) the distribution of probability of the slope modulus ξ_m corresponds to the distribution of the random value modulus distributed according to the normal law with the mathematical expectation equal to zero:

$$P_{mG}(\xi_m) = \frac{2}{\sqrt{2\pi}\,\sigma} \cdot \exp\left(-\frac{\xi_m^2}{2\sigma^2}\right) \tag{9.12}$$

where parameter σ^2 is related to the variance of the slope module $\overline{\xi_m^2}$ by the ratio: $\overline{\xi_m^2} = 0.3634 \cdot \sigma^2$.

From (9.12) we get the ratio linking τ_Σ / T and σ,

$$\frac{\tau_\Sigma}{T} = \int\limits_0^{\xi_K} p_{mG}(\xi_m) \cdot d\xi_m \tag{9.13}$$

If the surface wave field is isotropic, $\gamma = 1$ and the parameter ξ_m is subject to the Relay distribution. The probability density function of the slope module can be represented as follows

$$P_R(\xi_m) = \frac{\xi_m}{\overline{\xi_0^2}} \cdot \exp\left[-\frac{1}{2} \cdot \frac{\xi_m^2}{\overline{\xi_0^2}}\right] \tag{9.14}$$

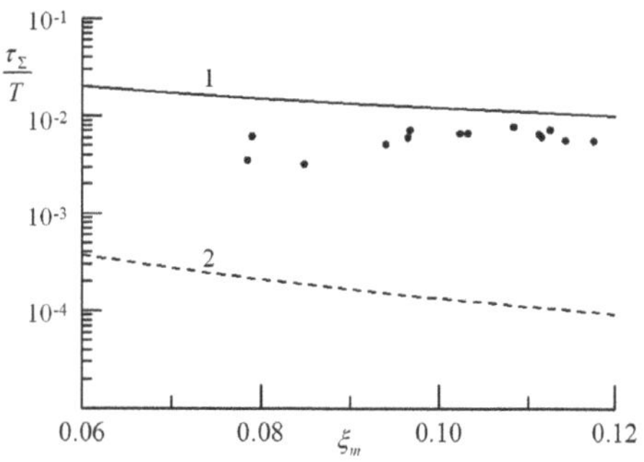

Fig. 9.2 Dependence of the probability of the laser beam reflected from the sea surface hitting the photoreceiver aperture τ_Σ / T from the slope module ξ_m. Curves *1*, *2*—calculation by Eqs. (9.13) and (9.17) respectively; points—experimental data

The variance of the slope module $\overline{\xi_m^2}$ is related to the ratio ξ_0 parameter

$$\overline{\xi_m^2} = \left(2 - \frac{\pi}{2}\right)\overline{\xi_0^2} \tag{9.15}$$

The probability that the slope module is less ξ_K defined as

$$\Phi_R(\xi_K) = 1 - \exp\left[-\frac{1}{2} \cdot \frac{\xi_K^2}{\overline{\xi_0^2}}\right] \tag{9.16}$$

We get

$$\frac{\tau_\Sigma}{T} = \Phi_R(\xi_K) = 1 - \exp\left(-\frac{0.215\,\xi_K^2}{\overline{\xi_m^2}}\right) \tag{9.17}$$

The results of calculations by Eqs. (9.13) and (9.17), as well as experimental estimates obtained from data of synchronous measurements and laser sounding are presented in Fig. 9.2. Model calculations have been carried out for $\xi_K = 0.0025$, that corresponds to conditions of carrying out measurements on an oceanographic platform at height of lidar height equal to 5 m.

The values determined in the experiments τ_Σ / T are placed closer to curve 1 corresponding to the situation when $\gamma = 0$, than curve 2 corresponding to the situation when $\gamma = 1$. This is somewhat unexpected, since the values of the three-dimensional index γ are mostly between 0.65 and 0.85 [42]. This arrangement of experimental values τ_Σ / T with respect to curves 1 and 2 is likely due to deviations of slope component distributions from the Gauss distribution not accounted for in Eqs. (9.13) and (9.17).

Big differences between the model curves constructed for the situations when they $\gamma = 0$ and $\gamma = 1$ are caused by the peculiarities of behavior of the probability density functions $P_{mG}(\xi_m)$ and $P_R(\xi_m)$ at small values ξ_m. Small values of the sea surface slope module ξ_m, at which the laser glint hits the aperture of the photoreceiver,

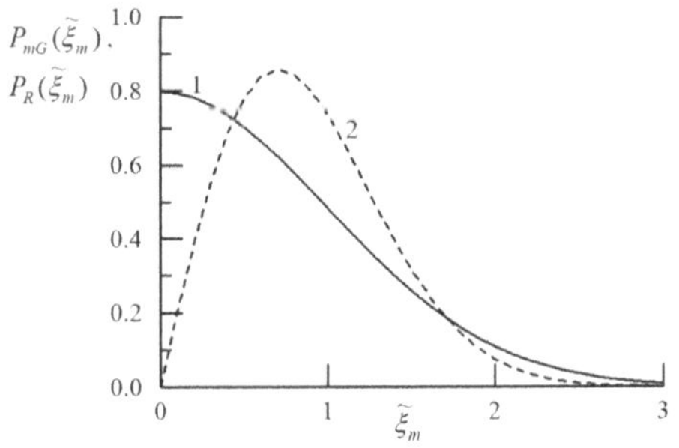

Fig. 9.3 Probability density functions built for $\gamma = 0$ (curve *1*) and $\gamma = 1$ (curve 2)

Table 9.1 Correlation matrix of glint reflectance and sea surface slope module parameters

	τ_Σ/T	F	I	τ
$\sqrt{\overline{\xi_m^2}}$	0.55	0.82	−0.61	−0.79
$\overline{\xi_m}$	0.51	0.80	−0.59	−0.87

have the highest probability density at, $\gamma = 0$ and at $\gamma = 1$ (Relay distribution) the probability density is close to zero (see Fig. 9.3).

Approaches used for interpretation of remote sensing data can be divided into two groups. The first group includes methods, in which a given parameter is determined within the model, built on the existing concepts of process physics. An example of such approach has been considered above. The second approach is to replace the model with a "black box" and the statistical connection between the parameters at its input and output is determined, i.e. directly between the measured parameter and the signal of the probing device.

Table 9.1 shows the correlation coefficients describing the relation between the statistical characteristics of highlights determined by vertical laser probing of the sea surface and its slope characteristics. The following laser glint characteristics are considered: τ_Σ/T is registration probability; F is registration frequency; I and τ is average intensity and duration of glint, as well as two parameters characterizing the slope module: average value and $\overline{\xi_m}$ standard deviation $\sqrt{\overline{\xi_m^2}}$.

It follows from Table 9.1 that the lowest correlation is observed between $\sqrt{\overline{\xi_m^2}}$, $\overline{\xi_m}$ and τ_Σ/T. The reasons for this have been discussed above. There is also a low level of correlation between the slope module parameters and the average intensity I. This is explained by the fact that the glint intensity is determined not only by the variance of the sea surface slope module but also by its curvature [40].

The high level of correlation of the slope module characteristics with the parameters F and τ allows using them to define $\sqrt{\overline{\xi_m^2}}$ and $\overline{\xi_m}$. The dependence of the parameters $\sqrt{\overline{\xi_m^2}}$ and $\overline{\xi_m}$ is also well F approximated by the linear regression equation (see Fig. 9.4).

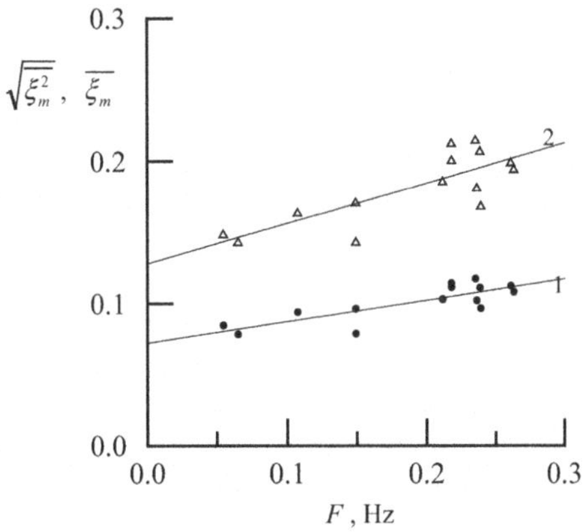

Fig. 9.4 Experimental dependencies $\sqrt{\overline{\xi_m^2}}$ (points) and $\overline{\xi_m}$ (triangles) on glint frequency *F*. Straight lines *1* and *2* are linear regressions (9.18) and (9.18) respectively

$$\sqrt{\overline{\xi_m^2}} = 0.072 + 0.150\,F \tag{9.18}$$

$$\overline{\xi_m} = 0.128 + 0.284\,F \tag{9.19}$$

The dependencies $\sqrt{\overline{\xi_m^2}}$ and $\overline{\xi_m}$ from τ are not linear (see Fig. 9.5). A polynomial of the second degree can be used to approximate them.

$$\sqrt{\overline{\xi_m^2}} = 0.190 - 0.039\,\tau + 0.0036\,\tau^2 \tag{9.20}$$

$$\overline{\xi_m} = 0361 - 0.079\,\tau + 0.0074\,\tau^2 \tag{9.21}$$

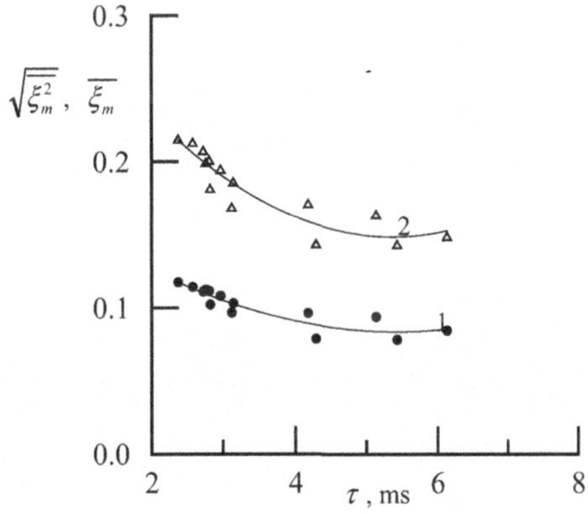

Fig. 9.5 Experimental dependencies $\sqrt{\overline{\xi_m^2}}$ (points) and $\overline{\xi_m}$ (triangles) on the glint duration *F*. Curves *1* and *2* are regressions (9.20) and (9.21) respectively

9.4 Statistical Connection of Laser Glint Characteristics and Wave Field Characteristics

The variability of the mirror reflection glint registered under different conditions is shown in Fig. 9.6, where the dependencies of the glint frequency and their average duration on the wind speed are given [41]. Due to the complexity of the graphical representation of a large number of measurements (for the analysis we used the data of 17,800 measurement sessions with the duration of 30 s each) the initial estimates of the frequency F and τ average duration of the glint were preliminary averaged by time intervals equal to 5 min, which corresponds to the averaging of 10 points.

The charts presented in Fig. 9.6 are based on data obtained in various hydrometeorological situations. Measurements were made at the stages of wave development

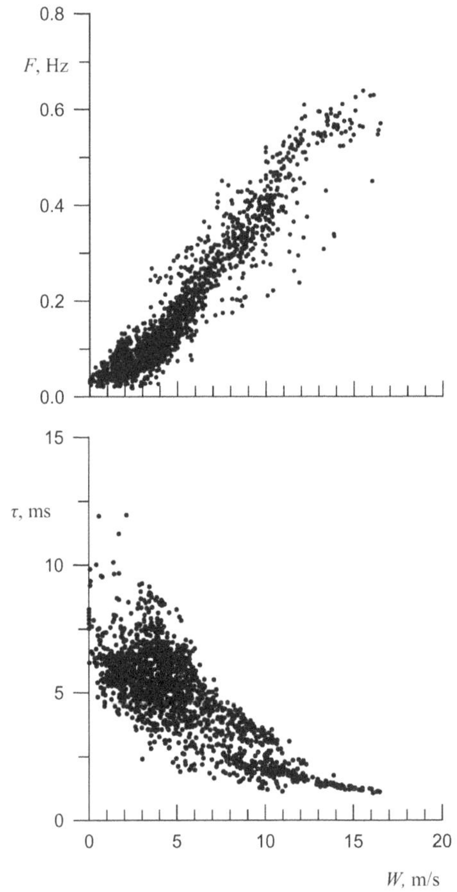

Fig. 9.6 Dependences of the frequency F and the average duration τ of glint reflection on the wind speed W

Table 9.2 Correlation matrix of statistical moments of longitudinal and transverse components of slopes

	$\overline{\xi_u^2}$	A_u	E_u	$\overline{\xi_c^2}$	A_c	E_c
$\overline{\xi_u^2}$	1	−0.368	−0.377	0.740	0.038	0.049
A_u	−0.368	1	0.128	−0.278	−0.008	0.049
E_u	−0.377	0.128	1	−0.230	0.020	0.059
$\overline{\xi_c^2}$	0.740	−0.278	−0.230	1	−0.340	−0.113
A_c	0.038	−0.008	0.020	−0.340	1	0.175
E_c	0.049	0.049	0.059	−0.113	0.175	1

and attenuation, under stationary conditions; at different fetch. A short fetch corresponded to a situation where the wind blew from the shore, a large fetch when the wind blew from the open sea.

The data presented in Fig. 9.6 confirm that the relationship between wind speed and parameters of the glint sequences is not deterministic, but has a stochastic character. This conclusion is in good agreement with the results of studies of the sea surface slope variability and its relation to wind speed [15].

It should also be taken into account that the statistical moments of different orders of the same component of the sea surface slopes are poorly correlated with each other (see Table 9.2). Accordingly, different values of the senior statistical moments may correspond to the same variance value of the slope component. As a consequence, at the same variance, there will be different deviations in the slope distribution from the Gaussian distribution. Besides, statistical moments of the same order of longitudinal and transverse slope components are poorly correlated with each other.

Let's build empirical dependences of the analyzed sequences of glint parameters on wind speed $F = F(W)$, $\tau = \tau(W)$ and $I = I(W)$. For this purpose, let's break the whole interval of wind speed changes into ranges of 1 m/s width at $W \leq 10$ m/s and 2 m/s width at $W > 10$ m/s. For each range, the average values of the analyzed parameters were calculated. Let us also calculate the standard deviations $\overline{\left(F-\overline{F}\right)^2}$, $\overline{(\tau-\overline{\tau})^2}$ and $\overline{\left(I-\overline{I}\right)^2}$. The type of curves describing changes in the mean values for the selected wind speed ranges, frequency $\overline{F}$ and duration $\overline{\tau}$ of glint, as well as their RMS is shown in Fig. 9.7. Note that the dependencies $\overline{F} = \overline{F}(W)$, $\overline{\tau} = \overline{\tau}(W)$ and $\overline{I} = \overline{I}(W)$ should be considered as a result of averaging on the ensemble of situations corresponding to different hydrometeorological conditions (different stages of wave development, different fetch, etc.).

From the graphs presented in Fig. 9.7 it follows that the dependence is not $\overline{F} = \overline{F}(W)$ monotonous. The function $\overline{F}(W)$ reaches its maximum at wind speed 14 m/s $< W <$ 16 W m/s. In this case the maximum of the function $\overline{F}(W)$ is near the edge of the range of wind speeds registered during the experiment. Note that a similar behavior of the function $\overline{F}(W)$ (presence of the maximum) was observed in another earlier experiment on an oceanographic platform in which the lidar was set at 12 m.

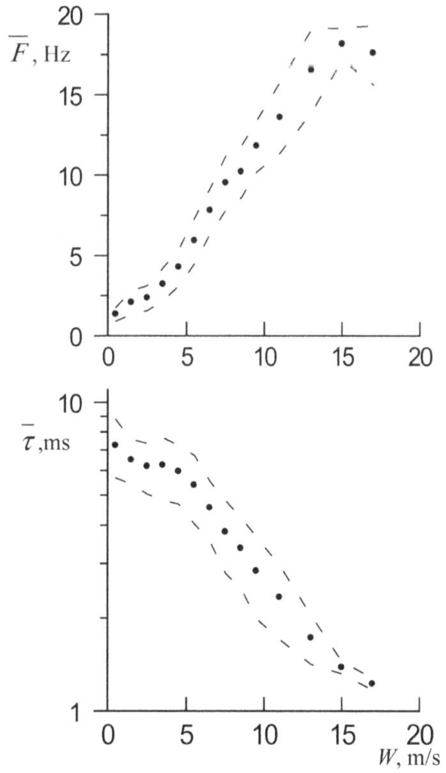

Fig. 9.7 Dependence of the average characteristics of glint reflectance $\overline{F}$ and $\overline{\tau}$ from wind speed W

During the measurement period, the weather conditions varied from calm to storm, with wind speed reaching 20 m/s. Maximum glint frequency values were recorded at wind speeds of $W \sim 10$–12 m/s.

To explain this effect, we can use spectral models describing changes in the energy of short surface waves with increasing wind speed [7, 27]. These models show that as the wind speed increases, the short wave energy level nonlinearly increases with the wind speed, the higher the wind speed, the slower the energy growth. However, the change in glint frequency is different. In the range of 5 < W < m/s 13 m/s, the glint frequency increase $\overline{F}(W)$ is almost linear.

We will compare the behavior of the function $\overline{F}(W)$ with the behavior of other characteristics determining the level of sea surface roughness at different wind speeds. These characteristics change monotonically with growth W. This is true for the sea surface slope variance determined from the data of satellite optical scanners [5] and from the data of laser sensing from spacecraft [13]. The local curvature of the sea surface grows nonlinearly monotonically with increasing wind intensity [2, 7].

These experimental results disprove the assumption made in [30] that saturation occurs at high wind speeds. The behavior of the function $\overline{F}(W)$ in the region of strong winds cannot be explained by changing any one characteristic of the wave field.

Similar dependence of glint frequency on wind speed as shown in Fig. 9.7 was obtained in [38]. In this work the possibility of wind velocity determination from the data of laser glint density measurements from a fast moving platform was considered, which corresponds to the condition of "frozen" sea surface. It was suggested that the nonmonotonic dependence of the average frequency of the mirror points $\overline{F}(W)$ on the wind speed can be a consequence of the competition of two physical processes. As the wind intensifies, the energy of short waves increases, which determine the level of roughness of the sea surface. This leads, in particular, to the growth of values of parameter F. Simultaneously with the roughness growth, the surface curvature at the mirror reflection points increases. Starting from a certain moment, a part of glint, though falling into the aperture of the photoreceiver, is not recorded, because their intensity is below its sensitivity threshold. At further growth of local curvature the number of glint with intensity below the threshold increases and, at excess of wind speed of some value W_0, dependence of frequency of glint registration on wind speed changes its character, namely, with growth of wind speed frequency F decreases.

The wind speed W_0 threshold entered here is determined both by the characteristics of the sea surface and by the own characteristics of the laser probe. It depends on such parameters as the power of the transmitter and the sensitivity of the photoreceiver as well as the distance between the instrument and the reflective surface.

The explanation of the behavior of the function $\overline{F}(W)$ proposed in [38] agrees well with the data from the experiments on a stationary oceanographic platform, where a decrease in the threshold value of wind speed W_0 when the lidar is removed from the sea surface was observed [41]. As the distance between lidar and the irradiated surface increases, the intensity of the reflected signal decreases, the sensitivity threshold is reached at lower wind speeds.

The intensity of mirror reflection I glint when probing the water surface is usually associated with the surface curvature at the point of mirror reflection [19, 30]. The value of parameter I is determined by the expression

$$I = \vartheta \, |K|^{-1} \tag{9.22}$$

where ϑ—proportionality factor depending on the characteristics of the measuring equipment; K is Gaussian curvature module.

The parameter $K = (r_1 r_2)^{-1}$ depends on the main radii of curvature r_1 and r_2. Their product is a measure of the local curvature of the surface. Usually, at carrying out of experimental researches of curvature of a surface, the factor ϑ was defined on the basis of calibration of intensity of the reflected laser signal from mirror spherical surfaces with the set radii of curvature [19]. Sometimes, short-period waves are considered as "quasi-dimensional", i.e., as waves that can be modeled in the form of cylinders. In this case, calibration of equipment for laser sensing of the sea surface to determine the curvature is performed on mirror cylinders with known radii.

It should be noted that the experiments in which the local curvature of the excited sea surface was determined by two independent methods: the method of laser sounding and with the help of other independent control equipment were not conducted. The question of what model is better to use to describe the local curvature of the real wave field, as well as the transition from measurements of the intensity of reflected laser pulses to the calculation of the main radii of the curvature, can not be considered solved. One of the disadvantages of such a technique is that two or more mirror points may be simultaneously present in a laser spot.

Two approaches are used to analyse the intensity of laser glint. In the first approach, when calculating the average intensity, each pulse has the same weight and its maximum value is determined. In the second approach, which is implemented here, averaging takes into account the pulse duration and its amplitude change over time.

The intensity of radiation reflected from the sea surface is calculated in the Kirchhoff approximation. The conditions, the length of the electromagnetic wave is small in comparison with the radius of curvature of the surface, is always fulfilled at laser probing. The analysis is reduced to determining the characteristics of the surface glint reflectance points. Experimental studies have shown that the intensity of glint reflectance I ambiguously depends on the wind speed. The obtained estimates of I are shown in Fig. 4.4. The values of parameter I are given in conventional units. Changes of parameter I with growth of wind speed occur non-monotonically, is observed minimum at wind speeds of 7–8 m/s. At $W < 8$ m/s approximation is fair.

$$I \sim W^{-0.48} \tag{9.23}$$

It is necessary to explain not monotony of changes of average intensity of glint at growth of wind speed and occurrence of minimum at $W = 7–8$ m/s. This may be due to the action of several physical mechanisms that determine the surface roughness. One of such mechanisms is the collapse of waves. At wave crest collapse, spots of smoothed surface are formed, which are areas with low local curvature. Their appearance is manifested in the form of impulses with high intensity, which, perhaps, is one of the reasons observed in Fig. 9.8. Some growth of intensity at $W > 8$ m/s. Surface wave crest collapses also play an important role in the formation of the radio signal scattered by the sea surface [31].

When the laser beam is reflected from the area of the smoothed surface, not only the amplitude of the recorded pulse increases, but also its duration. Large radii of surface curvature correspond to large time intervals during which the condition of mirror reflection is realized (9.3). As a result, a single reflection from a section of the smoothed surface has a greater weight in estimating the average intensity per session than a single reflection from a rough surface.

Another hydrodynamic mechanism is the Langmuir circulation that occurs when the wind increases in the surface layer of the sea, which forms a system of alternating zones of convergence and divergence. Like the collapse of waves, the appearance of Langmuir cells leads to the appearance of zones on the surface where the ripple waves are suppressed. Langmuir cells form long stretches of creamy stripes along the wind direction and are clearly visible at wind speeds above 5 m/s.

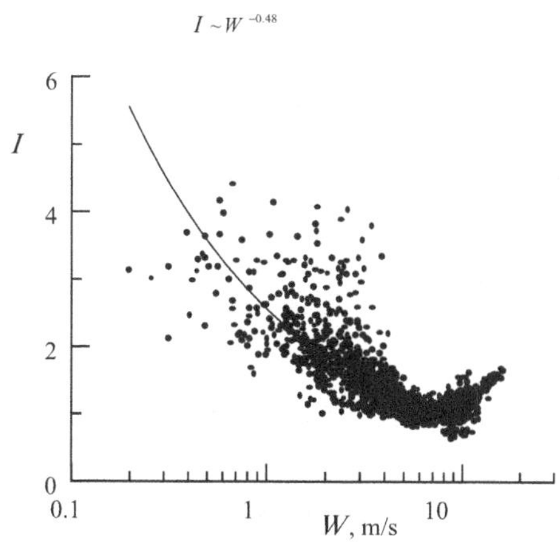

Fig. 9.8 Dependence of average reflection intensity I on wind speed W; points is measurement data, line is regression dependence, constructed for wind speeds less than 8 m/s. The values of parameter I are given in conventional units

Among the analyzed functions, the only function $\overline{\tau}(W)$ that changes monotonically with the wind speed is the function that allows to use it for remote determination of wind speed from laser sensing data. However, taking into account that the nature of the relationship between laser glint parameters and wind speed is stochastic, and given the different sensitivity of these parameters to changes in wind speed, it is reasonable to build biparametric models of the type $W = w(F, \tau)$.

The most sensitive parameter to changes in wind speed is the average frequency of recorded glint $\overline{F}$. Relative sensitivity, which we define here using the dynamic range of changes in the investigated parameter. Let's accept, that relative sensitivity is equal to the relation of the maximum value of investigated parameter to its minimum value, for it $\overline{F}$ is equal 13.4. The sensitivity $\overline{\tau}$ calculated by similar method, $\overline{I}$ accordingly, is equal to 4.9 and 2.9.

Laser Glint Indication in the Areas of Slick-Ripple

Changes in the glint characteristics when the laser beam hits the "smoothed" areas of the surface have been studied in two major slick situations frequently encountered in the area of the oceanographic platform. These are slicks created by internal waves and Langmuir circulation, which are easily distinguished by visual observations.

Changes in the nature of the reflected signal when passing the slick created by the internal waves is shown in Fig. 9.9. Measurements were made in the summer period with stable wind direction and speed. The average wind speed was 5 m/s. To estimate the relationship between the glint characteristics and the intervals of slicks appearance, the latter were presented in the form of a binary series consisting of zeros and units, in which zeros correspond to the slice and units to ripples.

As can be seen from Fig. 9.9, in the zone of slick created by internal waves, there is a halving of the glint frequency, from 10–12 Hz to 5–6 Hz; the intensity of the reflected signal increases by 4–8 dB and there is a small (about 20%) decrease in the duration of glint.

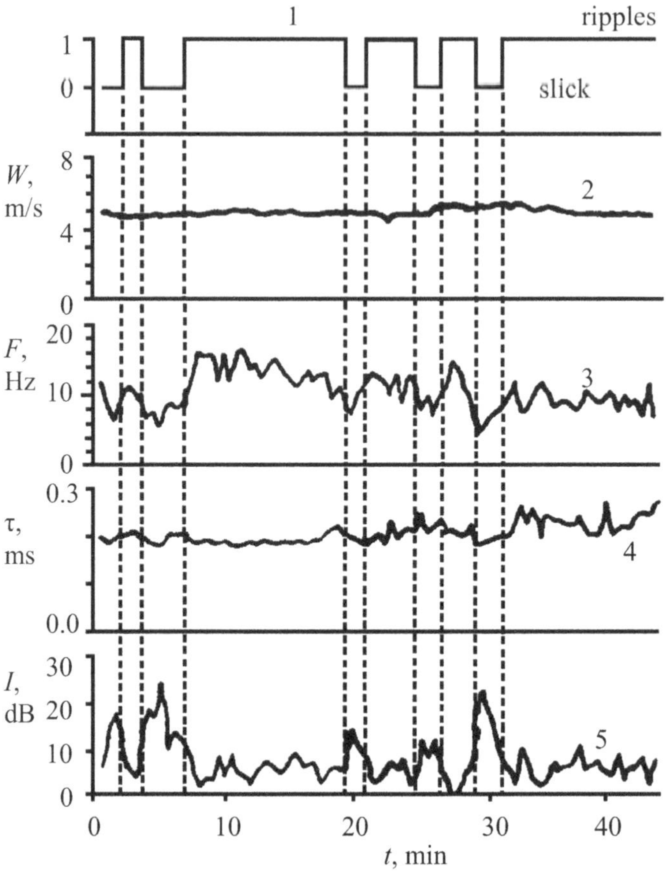

Fig. 9.9 Laser signal variations in areas of slicks created by internal waves: curve 1 is schematic image of slicks and ripples; curve 2 is wind speed; curve 3 is glint frequency *F;* curve 4 is duration τ; curve 5 is intensity *I*

Let's compare the change in glint characteristics in slicks created by different physical mechanisms. Changes in slicks formed by Langmuir circulation are shown in Fig. 9.10.

In both situations, when making measurements in slicks created by internal waves and Langmuir circulation, the signal of the device for laser probing, as well as meteorological parameters was recorded with an average of 30 s. The uniform measurement technique allows comparing effects created by different mechanisms.

Comparing Figs. 9.9 and 9.10, it can be noted that in Langmuir slicks variations in intensity and duration of glint have the same order as in slicks created by internal waves. Changes in *F* frequency are slightly lower than in the passing of internal waves. The significant difference in average values of glint parameters is obviously due to different weather conditions. Slicks from internal waves are usually observed at

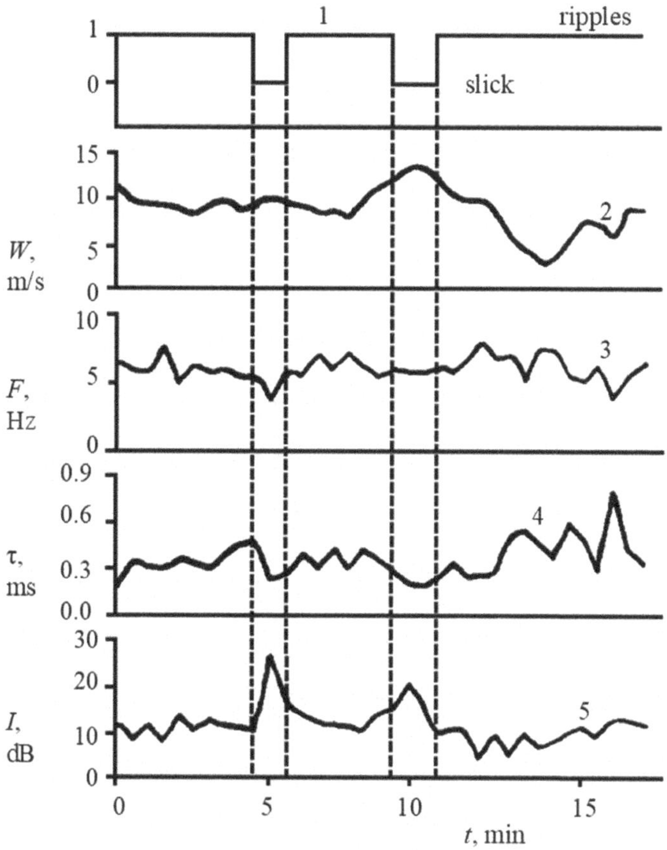

Fig. 9.10 Variations of laser signal in zones of slicks created by Langmuir circulation: curve 1 is schematic representation of slicks and ripples; curve 2 is wind speed; curve 3 is glint frequency *F*; curve 4—duration τ; curve 5 is intensity *I*

wind speeds up to 5–6 m/s, when the wind increases these slicks disappear, instead of them appear slicks created by Langmuir circulation. Visually slicks created by internal waves and Langmuir circulation are clearly distinguishable.

A series of experiments to study the effects of surfactants on the sea surface were conducted on the oceanographic platform. A small amount of the solution (~300 ml) was poured out near the laser-lit zone, after which a reduced roughness zone appeared on the sea surface for a few seconds, which expanded rapidly. As a result of the transfer by the current, the stain of contamination fell into the lidar field of view. In contrast to natural slicks, artificial slicks caused by surfactants entering the sea surface tend to exhibit sharper changes in the field of high-frequency surface waves. The following changes in the characteristics of the recorded signal have been observed. The frequency of glint decreased by 3–6 times, their duration increased by 1.5–2 times, and the intensity increased by 10–14 dB.

The significant change in reflection characteristics in the artificial slick zone allows the use of lidars as a means of controlling marine surface pollution.

Indication of Laser Glint During Upwelling

Upwelling is an important oceanographic phenomenon. Nutrient-rich cold water rises to the surface of the ocean, creating high biological productivity, beneficial for the growth of phytoplankton and other primary producers. Upwelling zones account for about half of the global fish catch [34]. Upwelling can be seen in sea surface images from different types of equipment installed on satellites, optical scanners, altimeters, and synthetic aperture radars [3, 8, 11].

There are two main types of upwelling: large-scale (or climatic [10]), which include constantly observed upwelling phenomena off the western coasts of the continents—California, Peruvian, South African, etc., and local synoptic, formed both in the open ocean areas (over banks, seamounts and ridges) and in marginal and inland seas. In the Black Sea, local upwelling events are not uncommon and occur in some areas of the shelf zone [11, 28].

The data of natural measurements and laboratory studies show that with changes in the temperature of the sea surface layer the dependence of the roughness of the water-air interface on wind speed changes [4]. The effect is determined by several factors that determine energy transfer from airflow to waves. As the stratification of the atmospheric drive layer changes, the wind velocity profile changes its shape and the pulse flow from wind to waves changes accordingly. Stratification stability leads to reduction of sea surface roughness, while instability leads to its increase.

As temperature changes, the physical properties of seawater change [35]. The parameters affecting the sea surface roughness are the dynamic viscosity and surface tension. Dynamic viscosity is highly dependent on temperature, it changes more than 2 times when the temperature changes by 30 °C. As the temperature drops, the kinematic viscosity increases, resulting in a faster attenuation of gravity-capillary and capillary waves. The surface tension changes due to temperature changes do not have a noticeable effect on the roughness, and they can be neglected.

Thus, a decrease in the temperature of the sea surface layer should lead to a reduction in roughness, as evidenced, inter alia, by remote sensing data. Cold sea fronts in radio-range sensing are observed as areas of reduced radio-wave scattering, i.e. as areas of reduced roughness. The upwelling zones to the surface of cold deep waters also have the same view on radar images, and upwelling appears as a region of reduced scattering [8, 12].

The opposite situation was recorded in the Black Sea during the oceanographic platform research during local upwelling off the Southern coast of Crimea [43]. The vertical thermal structure of sea water was measured by a complex of hydrophysical equipment consisting of 3 thermohirlands. Each thermohilland included 9 distributed temperature sensors, the length of which was equal to 0.92 m. The accuracy of temperature measurement was 0.1 °C [16]. In the local upwelling zone, during the period when the surface layer temperature dropped by about 10 °C, the sea surface roughness level was higher than at the same wind speeds before upwelling.

The intensity of dynamic processes in the upper sea layer at different stages of upwelling varies. The upwelling-downwelling cycle consists of three phases: the first one-cold water rise to the surface; the second one-isothermal phase; the third one-final phase when cold water is replaced by warm water. The highest intensity of dynamic processes is observed in the third stage, when cold water drops. At this stage isolated internal waves with amplitudes to 5–7 m are registered. At the stage of rise of deep waters courgas of internal waves with amplitudes of 1 m are registered. The ascending speed of cold deep waters is approximately two times less than their descending speed. Dynamics of local upwelling on the shelf near the Southern coast of Crimea is more intensive than climatic ones, where vertical speeds are 1–2 orders of magnitude lower. (Khristoforov, Zapevalov, [18]. Characteristic changes in the depth of isotherms at different stages of upwelling are shown in Fig. 9.11.

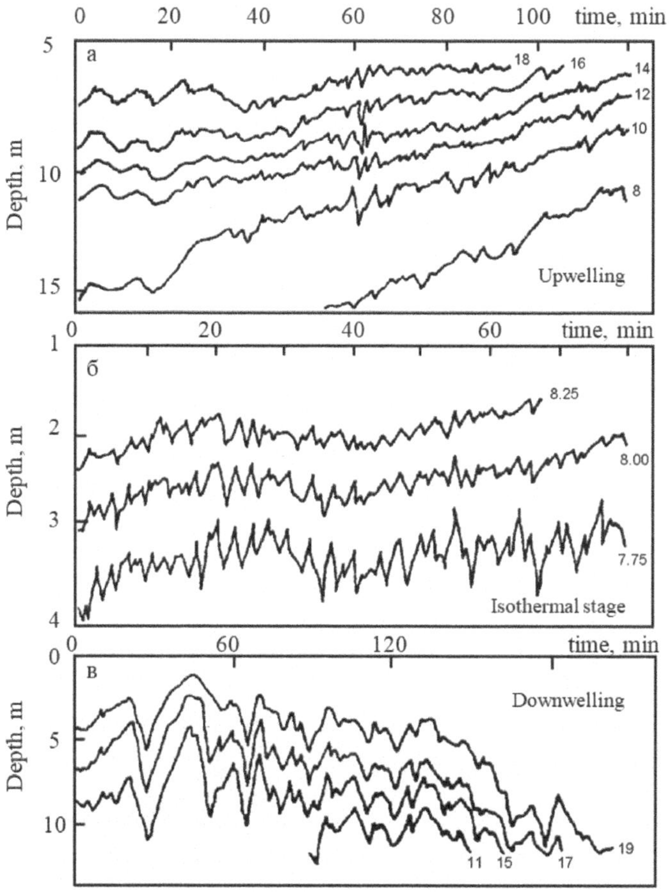

Fig. 9.11 Small-scale temperature field structure at different phases of the Black Sea local upwelling development. The numbers show the values of isotherms

In order to exclude the influence on the sea surface characteristics of intensive dynamic disturbances in the upper sea layer typical for the first and second phases of upwelling [18], here we limited ourselves to the analysis of the second isothermal upwelling phase. During this period, the water temperature at the surface approached 8.3 °C and at the bottom 7.5 °C, i.e., in the entire thickness from surface to bottom, a quasi isothermal stratification of temperature with a temperature gradient averaging 0.027 °C/m was observed. The surface roughness characteristics of the sea surface recorded during the second upwelling phase are compared with those obtained before the upwelling phase, when the temperature difference between water and air was small.

During the upwelling period, as shown by the results of laser sensing of the sea surface, the parameters characterizing the roughness level increase. In comparison with the estimates obtained in the period preceding upwelling, the frequency of mirror reflection glint registration during upwelling increases, and their average duration decreases (see Fig. 9.12).

Such a change in the characteristics of the lidar signal during the second upwelling phase corresponds to an amplification of roughness [17]. As will be shown below, this is in good agreement with measurements of the characteristics of short-period surface waves. For ease of comparison of data obtained under different conditions, the values of frequency and average duration of glint are given in Fig. 9.12 in normalized

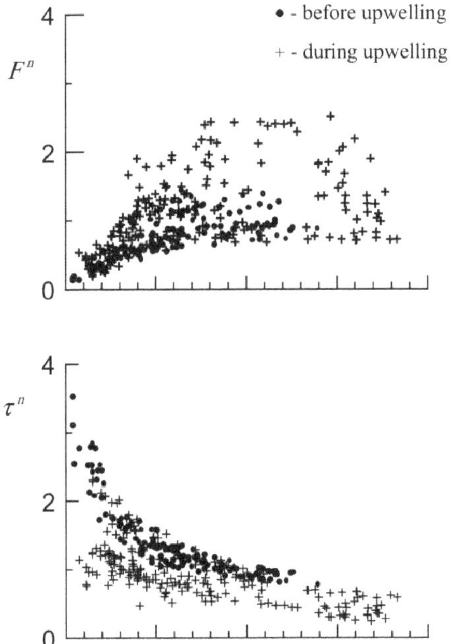

Fig. 9.12 Dependence of the estimates of the F^n mirror reflection frequency and their average duration τ^n on the wind speed W

form F^n and τ^n. Normalization was performed on their average values obtained from background measurements (before upwelling) at wind speed of 8–10 m/s.

Let's compare the roughness changes determined by laser probe data with the results of direct waveform measurements, which were made simultaneously. The measurements were carried out by a string breaker. RMS amplitudes of spectral components were determined by means of $\eta(f_i)$ parallel spectroanalyzer. The frequency range in which the signal is analyzed is 0.4–23 Hz. Figure 9.12 shows the dependencies of wind speed *W of the root* mean-square amplitudes of the spectral wave components measured during the second stage of fully developed upwelling as well as in the period before the upwelling.

The energy of the spectral components of the surface waves analyzed here differs by several orders of magnitude; therefore, for the convenience of comparison, a normalization in the form $\tilde{\eta}(f_i) = \frac{\eta(f_i)}{\langle\eta(f_i)\rangle}$ where the symbol $\langle\rangle$ denotes the average value in background measurements has been introduced. Average values $\eta(f_i)$ obtained before upwelling for wind speeds from 8 to 10 m/s were taken as background characteristics.

As can be seen from Fig. 9.13, the effects associated with upwelling appeared in wind waves with frequencies around 4.3 Hz and above. These effects are most pronounced in gravitational capillary waves and capillary waves. As the frequency of the waves increases f_i, the difference between the estimates obtained before and during upwelling increases.

Since a decrease in the sea surface temperature should lead to a decrease in the roughness level, and we see the opposite effect, other physical factors should be considered. Such factors may be specific hydrometeorological conditions. Let's consider the hydrometeorological conditions at which the roughness increase in the second phase of fully developed upwelling was observed. During the period under consideration the wind changed sharply in speed and direction. The wind from the shore (300°–320°) corresponded to high velocities, in this direction the distance from the shore to the platform was ~1 km. In this case, with a wind speed of 10 m/s before reaching the platform, the air is above the cold underlying surface for about 100 s. It is natural to assume that the boundary layer of the atmosphere, despite the large temperature difference between water and air, does not significantly change its characteristics during the specified time, and the effects associated with changing its stratification should be weak.

An important factor that can affect the roughness level of the sea surface is the current. A two-day fragment of the current velocity and direction recording is shown in Fig. 9.14. Measurements were made at a depth of 5 m, the measurement resolution was 30 min. During the second phase of upwelling, in Fig. 9.14 it is highlighted by arrows, the current velocity was small and did not exceed 9 cm/s. The measurement data presented in Fig. 9.14 indicate that the current could not have had a noticeable effect on the sea surface roughness level.

As a rule, upwelling zones are characterized by high biological productivity [34]. Bioimpurities by adsorption from the water column form a natural film of surfactants on the sea surface. The film changes the size of surface tension, it leads to suppression of the short-wave part of the wind wave spectrum. The film also changes the flow

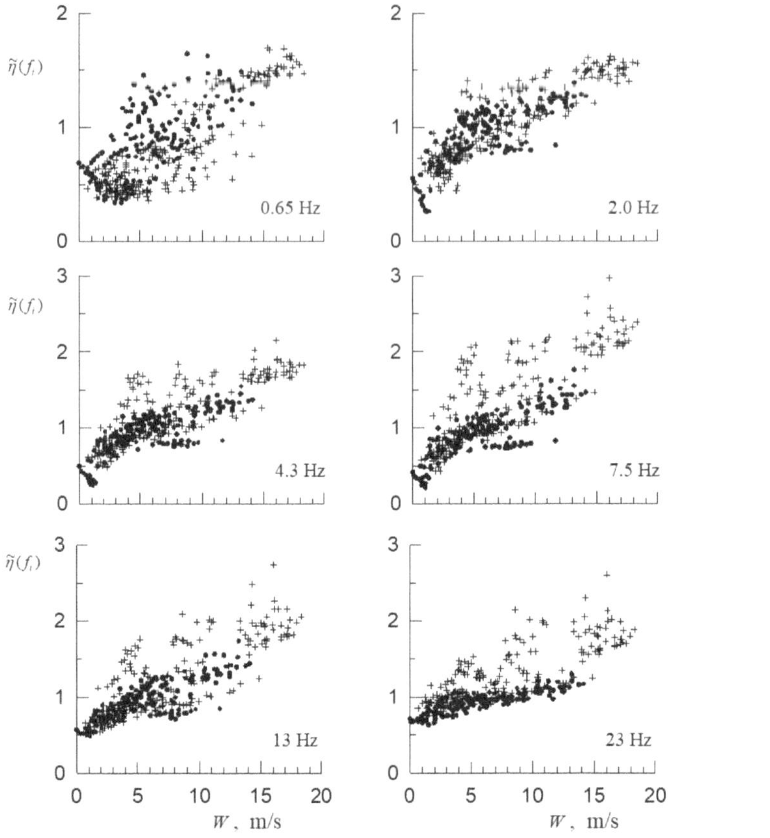

Fig. 9.13 Dependence of relative changes of mean square amplitude of wave components $\tilde{\eta}(f_i)$ (• is measurements before upwelling, + is measurements during upwelling) on wind speed W

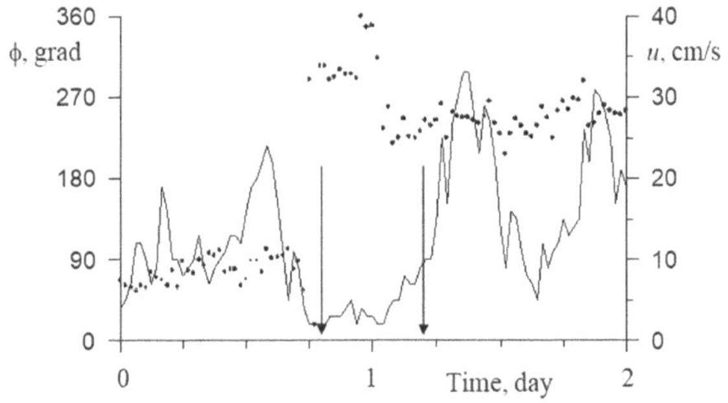

Fig. 9.14 Speed u (curve) and flow direction ϕ (dotted line) at 5 m horizon during upwelling period

of energy from wind to waves and modifies the nonlinear interactions between the wave field components.

In order to explain the observed increase in roughness level in the local Black Sea upwelling, it is necessary to identify significant characteristics that distinguish it from large-scale upwelling. Such characteristic may be the time of cold water stay near the surface. For Black Sea upwelling, the maximum frequency is about a day [28]. In the local upwelling considered here, cold waters rising from the depths were near the surface for a relatively short period of time. In the situation under consideration, the second (isothermal) phase of upwelling lasted about 10 h. Cold deep waters contain significantly less dissolved surfactants than warm surface waters in the coastal zone. During the short period of the second phase of local upwelling, despite the increase in biological activity, there is no time to form a film on the surface creating the same surface tension level as before upwelling.

It can be concluded that in local upwelling, when cold waters are at the surface for a relatively short period of time, there are two oppositely directed processes that affect the roughness of the sea surface. The change in atmospheric drive layer stratification as a result of a decrease in the temperature of the sea surface layer leads to a decrease in the roughness level and a decrease in the concentration of surfactants to its growth. The conducted studies indicate that a second process prevails in the considered local coastal upwelling on the Black Sea. Such a physical interpretation allows to explain why, according to the measurements by the string breaker, amplitude differences at a fixed wind speed before upwelling and during its second phase were observed in high-frequency waves. With increasing frequency, the effect of surface tension increases, so the relative change in amplitudes must increase.

Indication of Laser Glint During Rain

With low winds, rainfall significantly changes the roughness of the sea surface. At wind speed 0–2 m/s, as Fig. 9.15 shows, lidar signal characteristics change significantly. With the onset of rain frequency F increased from 1–2 to 15–20 Hz, duration T decreased by 10–20 times: and the intensity of glint decreased by 10–15 dB. The intensification of rain was accompanied by a further increase in frequency to 30 Hz and an even deeper drop in intensity (20 dB compared to the background). With its termination, the glint structure recovered to the background level prior to the onset of rain, indicating the dynamic origin of surface structure disturbances caused by rain droplets [17].

9.5 Conclusion

The results of the cycle of experimental studies of the lidar signal variability when probing the sea surface are analyzed. The researches were carried out on a stationary oceanographic platform installed in the Black Sea.

The impact of air flow on the sea surface is the dominant physical mechanism for forming its structure. The main elements of the sea surface forming reflections are

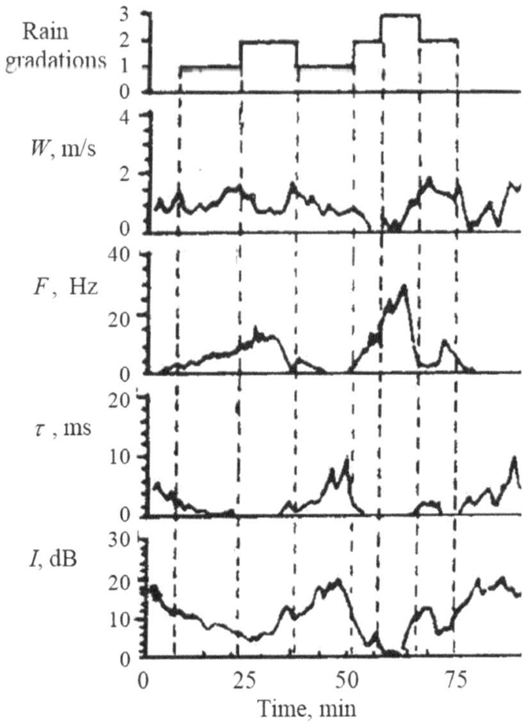

Fig. 9.15 Effects of rain on sea surface. Schematic representation of precipitation intensity gradations (level 1 is weak rain, level 2 is moderate, level 3 is strong)

short wind waves. Possessing relatively low energy, short waves change their characteristics under the influence of processes occurring in the atmosphere and thicker than the ocean, which allows to solve the reverse problem: to restore the characteristics of these processes. The statistical dependences describing the stochastic connection of the mirror reflection parameters with the wind speed are obtained. The application of the method of indication of laser reflections reflected from the excited sea surface makes it possible to reliably register structural anomalies appearing on it created by non-uniform and unsteady winds, non-uniform sea currents, internal waves and Langmuir circulation in the upper sea layer, anthropogenic pollution and atmospheric precipitation.

The peculiarities of short surface wave field formation in the area of local upwelling have been investigated. It is shown that in contrast to large scale upwelling, local upwelling can reduce the sea surface roughness level. This effect is the result of two oppositely directed processes affecting the sea surface roughness. Decrease of the sea surface temperature leads to decrease of the roughness level, decrease of surface-active substances concentration to its growth. The second process may prevail in local upwelling.

References

1. Aleksandrov AP, Legeza VP (1988) Laser meter of rough sea surface RMS slopes. Phys Oceanogr 6:51–56
2. Apel JR (1994) An improved model of the ocean surface wave vector spectrum and its effects on radar backscatter. J Geophys Res 99(C8):16269–16291
3. Askari F (2001) Multi-sensor remote sensing of eddy-induced upwelling in the southern coastal region of Sicily. Int J Remote Sens 22:2899–2910
4. Bortkovskii RS (1997) On the influence of water temperature on the ocean-surface state and transfer processes. Atmos Oceanic Phys 33(2):242–248
5. Bréon FM, Henriot N (2006) Spaceborne observations of ocean glint reflectance and modeling of wave slope distributions. J Geoph Res Oceans 111(C06005). https://doi.org/10.1029/2005jc003343
6. Bufton JL, Hoge FE, Swift RN (1983) Airborne measurements of laser backscatter from the ocean surface. Appl Optics 22(17):2603–2618. https://doi.org/10.1364/AO.22.002603
7. Cheng Y, Liu Y, Xu Q (2006) A new wind-wave spectrum model for deep water. Indian J Marine Sci 35(3):181–194
8. Clemente-Colón P (2004) Upwelling. In: Christopher RJ, Apel JR (eds). Synthetic aperture radar marine user's manual. U.S. Department of Commerce, Washington DC, pp 189–206
9. Cox C, Munk W (1954) Measurements of the roughness of the sea surface from photographs of the sun glitter//. J. Optical. Soc. America. 44(11):838–850
10. Fedorov KN (1983) Physical nature and structure of ocean fronts. Gidrometeoizdat, Leningrad, p 296
11. Goryachkin YuN (2018) Upwelling nearby the crimea western coast. Phys Oceanogr 25(5):380–389
12. Hsu M-K, Mitnic LM, Liu C-T (1995) Upwelling area northeast off Taiwan on ERS-1 SAR images. In: COSPAR colloquium: space remote sensing of subtropical oceans, September 12–16 1995, Taipei, Taiwan, vol 2, pp 15B4-1–15B4-8
13. Hu Y, Stamnes K, Vaughan M, Pelon J, Weimer C, Wu D, Cisewski M, Sun W, Yang P, Lin B, Omar A, Flittner D, Hostetler C, Trepte C, Winker D, Gibson G, Santa-Maria M (2008) Sea surface wind speed estimation from space-based lidar measurements. Atmos Chem Phys 8:3593–3601
14. Kay S, Hedley J, Lavender S (2009) Sun glint correction of high and low spatial resolution images of aquatic scenes: a review of methods for visible and near-infrared wavelengths. Remote Sens 1(4):697–730. https://doi.org/10.3390/rs1040697
15. Khristoforov GN, Zapevalov AS, Babii MV (1992) Measurements of sea-surface roughness parameters during the transition from calm to wind generated wave. Izvestiya Akademii Nauk Fizika Atmosfery i Okeana 28(4):424–431
16. Khristoforov GN, Zapevalov AS, Smolov VE (1998) Temporal variability of the vertical thermal structure in the vicinity of south Crimea. Phys Oceanogr 9(1):33–40
17. Khristoforov GN, Zapevalov AS, Smolov VE, Feldman YuR (1994) Lidar sensing of topographic inhomogeneities on the rough sea surface. Phys Oceanogr 5(6):461–468
18. Khristoforov GN, Zapevalov AS (1997) Development of summer upwelling off the South Crimea coast. Russ Meteorol Hydrol (7):41–46
19. Kodis RD (1966) A note on the theory of scattering from an irregular surface. IEEE Trans Antennas Propag AP-14:77–82
20. Kudryavtsev V, Yurovskaya M, Chapron B, Collard F, Donlon C (2017) Sun glitter imagery of ocean surface waves: Part 1. Directional spectrum retrieval and validation. J Geophys Res Oceans 122(2):1369–1383
21. Kukushkin AS, Parkhomenko AV (2015) Evaluation of applicability of the satellite data for studying suspended organic matter variability in the surface layer of the Black Sea. Sovremennyye problemy distantsionnogo zondirovaniya Zemli iz kosmosa 12(1):195–205
22. Kushnir VM, Stanichny SV (2007) Color index in the north-western black sea on the space color scanner data. Earth Obs Remote Sens 4:62–73

23. Kuzmin AV, Goryachkin YuA, Ermakov DM, Ermakov SA, Komarova NYu, Kuznetsov AS, Repina IA, Sadovskii IN, Smirnov MT, Sharkov EA, Chuharev AM (2009) Marine hydrographic platform "Katsiveli" as a subsatellite test site in the Black sea. Earth Obs Remote Sens 1:31–44
24. Lancaster RS, Spinhirne JD, Palm SP (2005) Laser pulse reflectance of the ocean surface from the GLAS satellite lidar. Geo Res Lett 32:L22S10. https://doi.org/10.1029/2005GL023732
25. Lebedev NE (2013) Opredelenie skorosti privodnogo vetra i stepeni zagryaznennosti morskoi poverkhnosti po izlucheniyu, registriruemomu sputnikovymi opticheskimi skanerami v zone solnechnogo blika (Determination of wind speed and degree of contamination of the sea surface by radiation measured by satellite optical scanners in the area of sun glint). Ekologicheskaya bezopasnost' pribrezhnoi i shel'fovoi zon i kompleksnoe ispol'zovanie resursov shel'fa 27:49–54
26. Lebedev NE, Aleskerova AA, Plotnikov EV (2016) The development of optical methods for sea surface slope measurement. Sovremennye problemy distantsionnogo zondirovaniya Zemli iz kosmosa. 13(3):136–149
27. Liu Y, Su M-Y, Yan X-H, Liu WT (2000) The mean-square slope of ocean surface waves and its effects on radar backscatter. J Atmos Oceanic Technol 17:1092–1105
28. Lomakin PD (2018) Upwelling in the Kerch strait and the adjacent waters of the Black sea based on the contact and satellite data. Phys Oceanogr 25(2):114–123
29. Longuet-Higgins MS (1957) The statistical analysis of a random, moving surface. Philosophical Transactions of the Royal Society A: Mathematical, Physical and Engineering Sciences, 249(966):321–387. https://doi.org/10.1098/rsta.1957.0002
30. Nosov VN, Pashin SYu (1990) Statistical characteristics of wind waves in the gravitational-capillary region of the spectrum. Atmos Oceanic Phys 26(11):1161–1169
31. Phillips OM (1988) Radar returns from the sea surface—Bragg scattering and breaking waves. J Phys Oceanogr 18:1063–1074
32. Pokazeev KV, Zapevalov AS, Lebedev KE (2017) Measurements of sea surface slopes by laser sensing from a space vehicle. Moscow Univ Phys Bulletin 72(4):410–414
33. Pustovoytenko VV, Lebedev NE (2015) Comparison of sea surface slope statistical moments obtained by means of optical scanners and laser inclinometers. Sovremennye Problemy Distantsionnogo Zondirovaniya Zemli iz Kosmosa 12(1):102–109
34. Ryther JH (1969) Upwelling. Science 166:72–76
35. Sharqawy MH, Lienhard JH, Zubair SM (2010) Thermophysical properties of seawater: a review of existing correlations and data. Desalin Water Treat 16(1–3):354–380. https://doi.org/10.5004/dwt.2010.1079
36. Shaw JA, Churnside JH (1997) Scanning-laser glint measurements of sea-surface slope statistics. Appl Opt 36(18):4202–4213
37. Suetin BS, Korolev SN, Suslin VV, Kucheryavy AA (2004) Manifestation of optical properties of atmospheric aerosol over the Black sea when interpreting data from the SeaWiFS satellite instrument. Phys Oceanogr 1:69–79
38. Vilchinskaya TB, Mikhailevich VG (1990) Using the method of laser sounding of the sea surface to determine the near-surface wind speed. Phys Oceanogr 6:57–61
39. Viollier M, Tanré D, Deschamps P-Y (1980) An algorithm for remote sensing of water color from space. Boundary Layer Meteor 18:247–267
40. Wu J (1971) Slope and curvature distributions of wind-disturbed water surface. J Opt Soc Am 61(7):852–858
41. Zapevalov AS (2000) Variability of lidar signal features when remotely sensing the sea surface. Oceanology 40(5):742–746
42. Zapevalov AS (2002) Statistical characteristics of the moduli of slopes of the sea surface. Phys Oceanogr 12(1):24–31
43. Zapevalov AS (2008) Statistical models of the sea surface in problems of acoustic and electromagnetic radiation scattering. In: Manuscript to claim the academic degree of doctor of physico-mathematical sciences on the speciality 04.00.22—geophysics. Marine Hydrophysical Institute, Sebastopol
44. Zapevalov AS, Pokazeev KV (2004) Sea surface slopes statistics and its application to the laser sounding. Vestnik Moskovskogo Universita. Ser. 3 Fizika Astronomiya 5:53–56

Chapter 10
Physical Limitations of Accuracy of Remote Determination of Wind Speed Over the Ocean

10.1 Introduction

For the first time, the idea of using a radar mounted on a spacecraft to measure drive wind speed was expressed in a 1965 paper at International Symposium, Electromagnetic Sensing of the Earth from Satellites [27]. It was implemented in 1973 at the orbital space station Skylab [28].

The idea of remote measurement of air flow over the sea is based on the dependence of the roughness level of the sea surface, which determines the power of the reflected radio signal from it, on the wind speed. Currently, the method of determining wind speed using remote sensing is the only way to obtain globally information about the wind above the sea surface. Two satellite systems for ocean wind estimation are the altimeter and the scatterometer.

A factor limiting the accuracy of remote determination of wind speed is that the relationship between the level of roughness and wind speed is not strictly determined [15, 20]. This connection is stochastic [21]. Although the wind is the main physical mechanism that determines the topographic structure of the sea surface, there are a number of other mechanisms that affect the characteristics of short surface waves.

Wind speed is determined by measuring the power received by the radar signal, which depends on the roughness of the sea surface and, therefore, on the wind speed. It should be noted that radar is a means of indirect measurement. Satellite radars are calibrated by comparing remote sensing data with measurements from oceanographic buoys [7, 14, 18]. Factors that reduce the accuracy of measuring wind parameters can be divided into three groups [37]. The first group includes technical factors, including various kinds of hardware noises. The second group includes errors that occur when calibrating satellite radars as wind speed meters. This group includes non-synchronization of measurements and deviations of the spacecraft path from the location of the buoy. The third group is the physical factors resulting from the

A. Zapevalov et al., *Simulation of the Sea Surface for Remote Sensing*, Springer Oceanography,
https://doi.org/10.1007/978-3-030-58752-9_10

ambiguity of the relationship between the characteristics of surface waves and wind speed.

In this chapter, we restrict ourselves to the analysis of the error caused by the third group, which determines the ultimate accuracy of measurements. In contrast to the standard approach, in which the measured data of wind speed from a spacecraft and a buoy are compared, here we examine the direct relationship between the characteristics of sea surface waves that determine the radio signal reflected from the sea surface and wind speed.

10.2 Altimetric Determination of Wind Speed

10.2.1 Radio Altimeter Signal Conditioning

In altimetry measurements from a spacecraft, the radio signal source and receiver are combined and are located far from the sea surface. In the geometric optics approximation, the specific effective area of scattering of radio waves when incident on a rough surface with finite conductivity is described by the expression [32]

$$\sigma_{sp} = \pi \sec^4 \theta |F_0|^2 P_{2D}(\xi_\uparrow, \xi_\perp)\big|_{\xi_\uparrow = tg\theta, \xi_\perp = 0} \tag{10.1}$$

where θ is incidence angle F_0 is Fresnel coefficient; P_{2D} is two-dimensional probability density function of slope; $\xi_\uparrow$ is slope of the sea surface in the sounding plane; $\xi_\perp$ is the slope of the sea surface in the orthogonal plane. In altimetry measurements, sounding is carried out in nadir, and we can take $\xi_\uparrow = 0$.

From Eq. (10.1) it follows that the error in determining wind speed from altimetry measurements depends on the nature of the relationship, the probability density function of the slope of the sea surface and wind speed. We will consider the nature of this connection based on the measurement data by a two-dimensional laser slopemeter and the sensing data of the sea surface in the optical range.

When constructing a model of specular reflection of radio waves by the sea surface, an approach based on approximations of geometric optics is used [3]. Specular reflection occurs if the condition is met: the surface wavelength is much greater than the length of the radio wave

$$\Lambda \gg \lambda \tag{10.2}$$

$$k\, R \cos^3 \theta \gg 1 \tag{10.3}$$

where Λ and λ is surface and radio wave lengths, respectively; $k = 2\pi/\lambda$ is wave number of the radio wave; R is local radius of curvature of the sea surface. Next, we will call long waves satisfying condition (10.2). Waves that do not satisfy conditions

(10.2) and (10.3) create diffuse scattering, which leads to a decrease in the power of the radar signal recorded on the satellite.

To take into account the phenomenon of diffuse scattering in expression (10.1), the Fresnel coefficient is replaced by the effective reflection coefficient. Its value depends on the wind speed. When the wind speed changes in the range of 0–25 m/s, the effective reflection coefficient varies from 0.38 to 0.5. If we narrow the range of variation of wind speed 1.5–15 m/s, then the value of the effective reflection coefficient changes by 10% (ranging from 0.45 to 0.50) [13].

Sea waves are a weakly nonlinear process [2]. The distribution of sea surface slopes is quasi-Gaussian [26]. Usually, the two-dimensional probability density function of the upwind ξ_c and crosswind ξ_c components of the slopes is described by the expression [4, 8]

$$P_{CM}(\xi_c, \xi_u) = P_G(\xi_c, \xi_u)\left[1 - \frac{1}{2}C_{21}H_2\left(\tilde{\xi}_c\right)H_1\left(\tilde{\xi}_u\right) - \frac{1}{6}C_{03}H_3\left(\tilde{\xi}_u\right)\right.$$
$$\left. + \frac{1}{24}C_{40}H_4\left(\tilde{\xi}_c\right) + \frac{1}{4}C_{22}H_2\left(\tilde{\xi}_c\right)H_2\left(\tilde{\xi}_u\right) + \frac{1}{24}C_{04}H_4\left(\tilde{\xi}_u\right)\right] \quad (10.4)$$

where

$$P_G(\xi_c, \xi_u) = \frac{1}{2\pi\sqrt{\overline{\xi_u^2}\,\overline{\xi_c^2}}}\exp\left(-\frac{\tilde{\xi}_c^2 + \tilde{\xi}_u^2}{2}\right) \quad (10.5)$$

probability density function Gaussian distribution $\tilde{\xi}_c = \xi_c/\sqrt{\overline{\xi_C^2}}$; $\tilde{\xi}_u = \xi_u/\sqrt{\overline{\xi_u^2}}$; $\overline{\xi_u^2}$ и $\overline{\xi_c^2}$—variances of the upwind and crosswind components of the slopes; H_n is n-th order Hermite polynomials; C_{ij} is coefficients. Hermite polynomials in Eq. (10.4) are described by the expressions

$$\begin{cases} H_1(x) = x \\ H_2(x) = x^2 - 1, \\ H_3(x) = x^3 - 3x, \\ H_4(x) = x^4 - 6x^2 + 3. \end{cases} \quad (10.6)$$

For definiteness, we assume that the first index of the coefficient C_{ij} corresponds to the crosswind component of the slope, the second corresponds upwind.

Equation (10.4) is obtained under the assumption that the distribution of slopes in the direction transverse to the wind direction is symmetrical. This assumption is not true in all situations. Significant deviations of the skewness of the crosswind component of the slopes from zero were revealed during experimental studies using a two-dimensional laser slopemeter on an oceanographic platform [22].

From Eq. (10.1) it follows that the error in determining wind speed from altimetry measurements depends on the nature of the relationship, the probability density function of the slopes of the sea surface and wind speed. We will consider the nature of this connection based on the measurement data by a two-dimensional laser slopemeter and the sensing data of the sea surface in the optical range.

10.2.2 Evaluation of the Error in Determining Wind Speed Based on in Situ Measurements

The error in determining wind speed due to the stochastic nature of the relationship between the slopes of the sea surface and wind speed U can be divided into two components. The first is the scatter of the variance of the slope components at fixed wind speeds $O_P(U)$, the second is the deviation of the distribution of the slope components from the Gaussian distribution. Accordingly, we represent the function O in the form of the product of two factors

$$O(U) = O_P(U)\,O_D(U) \tag{10.7}$$

where

$$\begin{aligned} O_P(U) = 1 &- \frac{1}{2}C_{21}H_2\left(\tilde{\xi}_c\right)H_1\left(\tilde{\xi}_u\right) - \frac{1}{6}C_{03}H_3\left(\tilde{\xi}_u\right) \\ &+ \frac{1}{24}C_{40}H_4\left(\tilde{\xi}_c\right) + \frac{1}{4}C_{22}H_2\left(\tilde{\xi}_c\right)H_2\left(\tilde{\xi}_u\right) \\ &+ \frac{1}{24}C_{04}H_4\left(\tilde{\xi}_u\right) \end{aligned} \tag{10.8}$$

$$O_D(U) = \frac{1}{\sqrt{\overline{\xi_u^2}\ \overline{\xi_c^2}}} \tag{10.9}$$

Comparison of the altimetric wind speed with the measurement data from buoys is usually carried out at small angles of incidence ($\theta \leq 0.017$ rad) [10]. The altimeter signal recorded on the spacecraft is formed by areas of the sea surface that satisfy the conditions $\xi_u \approx 0$ and $\xi_c \approx 0$. The polynomials H_1 and H_3 are odd functions of the slopes; therefore, in Eq. (10.4), the contribution of the terms proportional to these polynomials can be neglected. From Eq. (10.6) it follows that $H_2(0) = -1$, $H_4(0) = 3$, we obtain

$$P_{CM}(0,0) = P_G(0,0)\left[1 + \frac{C_{40}}{8} + \frac{C_{22}}{8} + \frac{C_{04}}{8}\right] \tag{10.10}$$

Coefficients C_{40}, C_{22} and C_{04} do not depend on wind speed. This means that deviations of the actual slope distributions from the Gaussian distribution are a factor leading to an error in altimetry measurements.

Further analysis is based on field measurements of sea surface slopes obtained using a two-dimensional laser slopemeter. The principle of operation of an optical device is based on measuring in two planes the angles of deviation of the laser beam from the vertical when an excited water-air interface passes from under water. The measurements were carried out on the oceanographic platform of the Marine Hydrophysical Institute, located in the coastal zone of the Black Sea. The depth at the installation site of the oceanographic platform is 30 m, the minimum distance to the coast is about 500 m [37].

To analyze the dependences of the values of the parameter O_D and O_P on the wind speed, presented in Fig. 10.1, the data of 276 measurement sessions were used.

As the wind speed increases, the variances of both slope components increase, and the parameter O_D values decrease accordingly. It can be seen that the relation between the parameter O_D and the velocity U is stochastic. Especially large scatter at wind speeds less than 4–5 m/s, with increasing speed it decreases. For the data

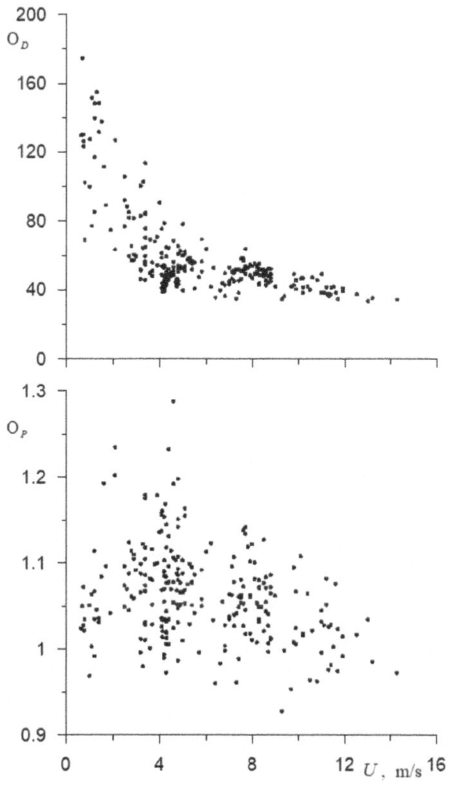

Fig. 10.1 Dependences of parameters O_D and O_P on wind speed U

presented in Fig. 10.1, the correlation between the parameter O_D and wind speed is −0.62.

Studies of the characteristics of the sea surface, which determine the level of roughness, show that at low wind speeds (up to 4–5 m/s) the variance of these characteristics is much greater than at higher wind speeds. Accordingly, a similar situation is observed when analyzing the variability of the parameter O_D. Note that at low wind speeds, the variance of slopes grows faster than at $U > 5$ m/s [36]

Values O_P vary between 0.9 and 1.3, which corresponds to a change σ_{sp} within 1.5 dB. The correlation between the parameters O_P and U is—0.30. We add that the parameters O_D and O_P are not correlated with each other.

Comparing the figures in Fig. 10.1 of the distribution of parameters O_D and O_D, we can conclude that the recorded dependence of the specific effective area of the altimeter scattering on the wind speed is mainly determined by the change in the variance of the components of the slopes. Deviations of the probability density function of slope components from the normal distribution, manifested in a significant scatter in the estimates of the fourth statistical moment, are a factor that reduces the accuracy of the altimetric determination of the drive wind.

It follows from expression (10.1) that, up to a constant factor, which is independent of wind speed, the function $\sigma_{sp}(U)$ must coincide with the function $O(U)$. Let us compare how the values O and σ_{sp} change at different wind speeds. For this, we compare the data obtained by measuring the slopes with the altimetry measurements of the GEOS-3 wind speed presented in Brown et al. [5].

As can be seen from Fig. 10.2, the behavior of parameters O and σ_{sp} the like. It is characteristic of them that in the region of weak winds they are very sensitive to changes in wind speed. At $W > 3$–4 m/s, sensitivity decreases markedly.

Estimates of the parameter O and wind speed U are related by the regression dependence

$$U = 1567 \exp(-0.3352\,\tilde{O}) \tag{10.11}$$

where $\tilde{O} = 10\,\lg(F)$.

The standard deviation of the U values from the approximation (10.11) turned out to be 1.73 m/s. The obtained deviation value almost coincides with the accuracy of a one-parameter determination of the wind speed from spacecraft [24].

Thus, the main characteristic determining the formation of the reflected radio signal during sounding from spacecraft to nadir is the sea surface slopes formed by short surface waves. The variety of factors affecting the energy level of the short-wave components of the wave field leads to an ambiguous dependence of the level of roughness of the sea surface on the wind speed. This circumstance limits the possible accuracy of the altimeter determination of wind near the sea surface.

It follows from the model of electromagnetic wave reflex by a sea surface during sounding into nadir that the accuracy of determining the wind speed is determined by two factors: a change in the type of distribution of the slopes of the sea surface and the variance of the slopes at fixed wind speeds. Changes in the kurtosis of the

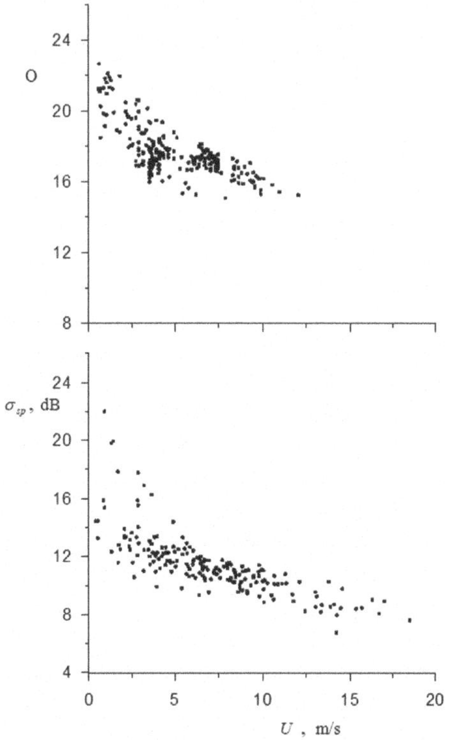

Fig. 10.2 Dependences of parameters O and σ_{sp} on wind speed U

components of the slopes are weakly correlated with the wind speed and lead to a spread of values σ_{sp} within 1.5 dB. The skewness of the slope distributions does not affect the value σ_{sp} during vertical sounding. The main factor that allows you to remotely determine the speed of the drive wind is the change in the variance of the slopes.

The scatter of the values of the slope characteristics obtained at the field experiment at a fixed wind speed should lead to a standard error of the altimetric determination of the wind speed of 1.73 m/s. A comparison of altimetry measurements of wind speed and measurements of wind speed from buoys showed that standard deviation of the difference is 1.74 m/s [5]. The value of this error practically coincides with the accuracy of the altimeter determination of wind obtained by comparing the data of altimeter measurements and measurements from meteorological buoys.

10.2.3 Estimation of the Error in Determining Wind Speed Based on Sounding Data of the Sea Surface in the Optical Range

In this work, for the quantitative analysis of the accuracy of determining the wind speed, we use the data of measurements of sea surface slopes and wind speeds presented in Bréon and Henriot [4], Cox and Munk [8]. In Bréon and Henriot [4], the statistical characteristics of sea surface slopes were calculated from optical images obtained using a POLDER scanner (POLarization and Directionality of the Earth Reflectances). Wind speed was determined using a NSCAT scatterometer (NASA Scatterometer). The scanner and scatterometer are installed on the ADEOS-1 spacecraft. In Cox and Munk [8], the characteristics of slopes were determined from aerial photographs of the sea surface in the flare zone. Wind speed was monitored by an anemometer mounted on the mast of the vessel. The distribution parameters (10.4) obtained in these works are presented in Table 10.1.

The regression dependences obtained by different methods describing the relationship between the variances of the slope components and the wind speed coincide up to the standard deviation (Fig. 10.3). However, the spread of statistical estimates of the slope variances obtained from aerial photographs is much larger (4–8 times) than according to satellite measurements. Also, at close average values of the coefficients C_{ij}, the standard deviation values of the SKO coefficients determined from aerial photographs are several times higher than the standard deviation values SKO obtained from satellite measurements.

To determine the error in measuring wind speed, taking into account Eq. (10.1), which relates the specific normalized sea surface radar backscattering cross-section with the probability density function of slope of the sea surface, we solve regarding variable ΔU the equation

$$P_{CM}(\xi_c, \xi_u, U + \Delta U) - P_{CM}^{\Delta}(\xi_c, \xi_u, U) = 0 \tag{10.12}$$

where P_{CM} is the probability density function (10.4), constructed with average values coefficients C_{ij}; P_{CM}^{Δ}—the same probability density function constructed with coefficients deviating from its average by the standard deviation. First, we determine

Table 10.1 Statistical characteristics of sea surface slopes

Bréon and Henriot [4]	Cox and Munk [8]
$\overline{\xi_u^2} = 0.001 + 0.00316\ U \pm 0.0005$	$\overline{\xi_u^2} = 0.000 + 0.00316\ U \pm 0.004$
$\overline{\xi_c^2} = 0.003 + 0.00185\ U \pm 0.0005$	$\overline{\xi_c^2} = 0.003 + 0.00192\ U \pm 0.002$
$C_{40} = 0.30 \pm 0.05$	$C_{40} = 0.40 \pm 0.23$
$C_{22} = 0.12 \pm 0.03$	$C_{22} = 0.12 \pm 0.06$
$C_{04} = 0.40 \pm 0.10$	$C_{04} = 0.23 \pm 0.41$

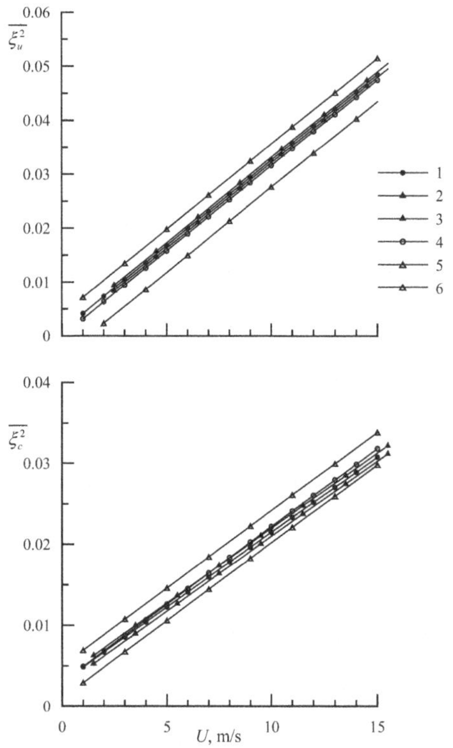

Fig. 10.3 Dependences of the variances of the upwind and crosswind components of slopes on wind speed U. Lines 1–3 are data [4], lines 4–6 are data [8], lines 1 and 4 are linear regressions, lines 2, 3 and 5, 6 are standard deviations from linear regressions

what errors the variance of the slope of the sea surface leads to at a fixed wind speed (error ΔU_V), then we evaluate the effect of deviations of the slope distributions from the Gaussian distribution (error ΔU_G). We assume that $\xi_c = \xi_u = 0$.

The results of calculating the absolute value of the error by the statistical characteristics of sea surface slopes presented in Bréon and Henriot [4], Cox and Munk [8] (they are indicated by $\Delta U_V^{(B)}$ and $\Delta U_V^{(C)}$, respectively) are presented in Fig. 10.4.

The calculations are performed for a situation when the variance of the slope components $\overline{\xi_u^2}$ and $\overline{\xi_c^2}$ simultaneously deviate from the average by the value of the standard deviation in the direction of higher or lower values.

The magnitude of the error $\Delta U_V^{(B)}$ is practically independent of the average wind speed. Given the classes of geophysical and oceanological problems that are currently being solved by remote sensing of the ocean, $\Delta U_V^{(B)} = 0.2$ m/s can be adopted. The error $\Delta U_V^{(C)}$ noticeably exceeds the error $\Delta U_V^{(B)}$. Depending on the average wind speed, the parameters $\Delta U_V^{(B)}$ and $\Delta\, U_V^{(C)}$ vary by 5–8 times.

The error due to the spread of the coefficients of the Gram-Charlier series C_{ij} is shown in Fig. 10.5. As in the analysis of the error ΔU_V, two cases were considered when the statistical characteristics were taken from Bréon and Henriot [4] (error $\Delta U_G^{(B)}$) and from Cox and Munk [8] (error $\Delta U_G^{(C)}$). It was assumed that the coefficients in expression (10.3) simultaneously deviate from their average values SKO by

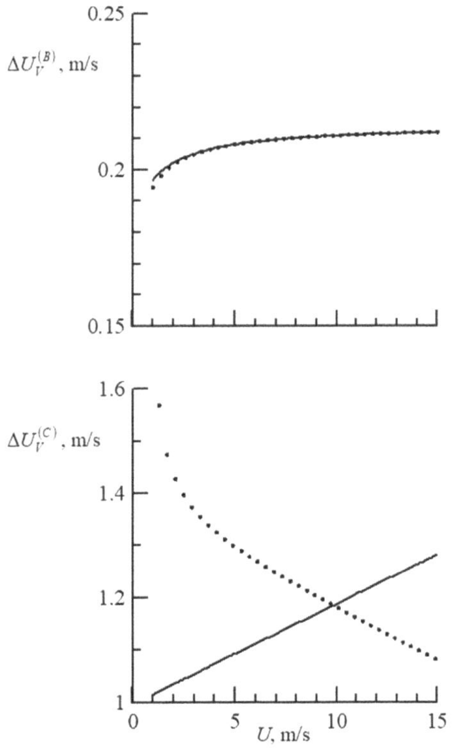

Fig. 10.4 Measurement error wind speed created ambiguity communication of energy of short waves and local wind. Solid line corresponds to large values of the variance of the slopes, dotted—less

the standard deviation in the direction of the values corresponding to the Gaussian distribution or in the other direction.

With increasing wind speed, the errors created by deviations of the distribution of slopes from the Gaussian distribution increase. The error $\Delta U_G^{(B)}$ is approximately four times less $\Delta U_G^{(C)}$.

From the analysis of Figs. 10.4 and 10.5 it follows that when comparing the characteristics of the slopes of the sea surface and wind speed, determined according to measurements from spacecraft, the error in determining the wind speed is much lower than the now accepted error in the altimetric measurement of the speed of the driving wind, which is 1.5–1.6 m/s [24]. At the same time, the error calculated by comparing the characteristics of the slopes determined from aerial photographs with direct measurements of the wind at a point (from the ship) is close to the altimetry error.

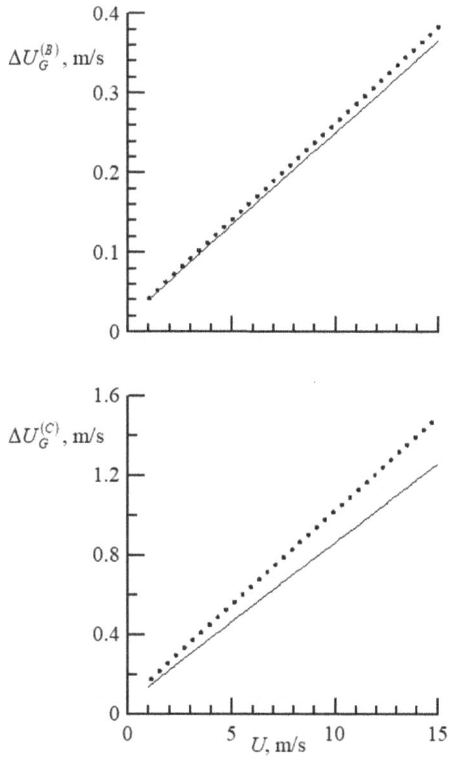

Fig. 10.5 Errors of measurement of wind speed $\Delta U_G^{(B)}$ and $\Delta U_G^{(C)}$ created by deviations distribution of slopes from the Gaussian distribution. Solid line corresponds large values of the coefficients given in Table 10.1, dashed—smaller

10.2.4 Two-Parameter Recovery of Altimetric Wind

The radio altimeter signal carries information not only about wind speed, but also about the height of energy-bearing surface waves. Analysis of the altimetry measurements of the GEOS-3 satellite showed that the steepness of the leading edge of the reflected pulse depends on the significant height of the wind waves [6, 33]. Currently, the standard product obtained by altimetry measurements is a significant wave height [16, 30]. A significant wave height is defined as the average height of 1/3 of the highest waves. It is associated with the variance of elevations of the sea surface $\overline{\eta^2}$ as

$$h_s = 4\sqrt{\overline{\eta^2}} \tag{10.13}$$

The creation of two-parameter algorithms, which take into account not only the backscattering cross section but also a significant wave height, can significantly (5–10%) improve the accuracy of restoration of the drive wind speed [18]. The magnitude of the correction for the state of the sea surface is calculated individually for each altimetry system (see, for example, [24] and the literature cited in the source).

Calibration of the altimeter as a wind speed meter is carried out on the basis of a comparison of the normalized backscattering cross section data and wind speed measurements obtained from measurements from buoys. In the same way, the effectiveness of wind speed calculation algorithms is estimated. It is usually assumed that a comparison can be made if the buoy and the point of altimetry measurements are separated in space by no more than 50 km, and the measurements are shifted in time by no more than 30 min, i.e. in this case, the sea surface can be considered as homogeneous and stationary [19].

In Chu et al. [7], studies were carried out on the fulfillment of the uniformity condition for sea-surface areas of different sizes. Studies conducted with the help of a rain locator showed that the indicated parameters are not always favorable for calibrating altimeters. It turned out that the condition of homogeneity within a circle of radius of 50 km was fulfilled in more than half of the measurements, and over 30% are inhomogeneous within a circle of radius of 20 km.

It should be noted that the backscattering cross section does not depend on the height of the waves, but on the variance created by the slopes of the sea surface. The correlation between the slopes and the significant wave height makes it possible to slightly increase the accuracy of wind speed recovery, but does not completely eliminate the ambiguity [19].

10.3 Scatterometer Determination of Wind Speed

As with altimetry measurements, the standard error between the values determined from the radar measurements and measurements from anchored buoys is used as an estimate of the accuracy of scatterometer determination of wind speed. At present, the accuracy of measuring wind speed with scatterometers is about 1.7 m/s, and the direction is 17 [11, 12, 20, 24, 31, 34]. This accuracy was achieved already in the first algorithms, where the backscattering cross section was used as the only predictor; for single-parameter models, it has not been improved so far. The accuracy of the measurement is determined by two factors: technical, including the non-synchronization of measurements described in the previous section, the mismatch of the location of the buoy with the satellite path, and physical, which consists in the correctness of the algorithm for calculating the wind speed from the backscattering cross section.

In the zeroth approximation, the scatterometer signal can be represented in the form [23]

$$\sigma_{pp}^{0} = 8\,\pi\,K^{4}\cos^{4}\theta\,\left|G_{pp}\left(\theta,\varepsilon\right)\right|^{2}\left[\Xi\left(\vec{k}_{B}\right) + \Xi\left(-\vec{k}_{B}\right)\right] \tag{10.14}$$

where pp is type of polarization, the first index corresponds to the polarization of the emitted signal, the second to the received; $\left|G_{pp}\left(\theta\right)\right|^{2}$ is the Bragg scattering geometric coefficient, which depends on the type of polarization of the radiated and received radar signal and on the dielectric constant of sea water; $\Xi\,(\vec{k}_B)$ is spectrum of sea surface waves, on the scale of the Bragg component. The resonance condition relates the surface wavelength λ_B to the radio wavelength Λ

$$\lambda_B = \Lambda / (2\,\sin\theta) \tag{10.15}$$

For vertical and horizontal polarization, the coefficient $G_{pp}(\theta, \varepsilon)$, respectively, has the form [32]

$$G_{vv}(\theta, \varepsilon) = \cos^2\theta \frac{(\varepsilon - 1)\left[\varepsilon\left(1 + \sin^2\theta\right) - \sin^2\theta\right]}{\left[\varepsilon\,\cos\theta + \sqrt{\varepsilon - \sin^2\theta}\right]^2} \tag{10.16}$$

$$G_{hh}(\theta, \varepsilon) = \cos^2\theta \frac{(\varepsilon - 1)}{\left[\cos\theta + \sqrt{\varepsilon - \sin^2\theta}\right]^2} \tag{10.17}$$

where indices v and h denote the vertical and horizontal polarizations, respectively.

Waves always exist on the sea surface, the length of which is much longer than the length of the Bragg components. Therefore, the Bragg components propagate along a curved surface created by long waves. Curvature of the surface leads to a change in the local angle of incidence. When irradiating such a surface, the local incidence angle is

$$\theta_L = \theta - \beta_\uparrow \tag{10.18}$$

where $\beta_\uparrow$ is the angle of the tilt of the sea surface created by long waves in the direction of sounding. Due to a change in the local angle θ_L, the resonance condition (10.15) changes, i.e. the wavelength at which scattering occurs changes, and the value of the geometric coefficient $\left|G_{pp}\,(\theta_L, \varepsilon)\right|^2$ also changes

The effect of the curvature of the surface along which the resonant waves propagate can be taken into account by averaging the backscattering cross sections σ^0_{pp} over the entire range of angles $\beta_\uparrow$. Averaging is carried out with a weight proportional to the probability density function $\beta_\uparrow$ [35]

$$\sigma^L_{pp} = \int \sigma^0_{pp}\left(\theta - \beta_\uparrow\right) P(\beta_\uparrow)\, d\beta_\uparrow \tag{10.19}$$

It follows from Eqs. (10.14) to (10.19) that the accuracy of the scatterometer determination of the wind speed depends on how the energy of the Bragg waves and the wind speed U are related and how $P(\beta_\uparrow)$ and U is related.

10.3.1 Short Wave Energy Versus Wind Speed

Most models connecting the level of spectral density of high-frequency waves with wind speed are based on theoretical analysis and laboratory measurements [1, 25]. Verification in natural conditions shows a significant scatter in their parameters, which imposes limitations on the accuracy of remote sensing estimates of the characteristics of the boundary layers of the atmosphere and the ocean [21].

Unlike laboratory conditions in real field conditions, the structure of high-frequency components of wind waves depends on a large number of various factors. Among them, one can point out the presence of surface-active substances of odd origin, technogenic pollution, surface currents and their local variations caused by the release of internal waves into the subsurface layer, vortex movements in Leigmuir circulations, orbital movements in swell waves, etc. High-frequency components of the wind disturbances, whose energy is 6–9 orders of magnitude lower than the energy of the energy-bearing wind will and swell, are largely affected by these factors. As a result, the structure of wind ripples (including its energy characteristics, on which the backscattering coefficient of radio waves depends) is not a reliable indicator of meteorological parameters in the drive layer of the atmosphere, as could be expected from laboratory experiments. Based on the foregoing, it is necessary to pay attention to the study of the actually observed relationships between the characteristics of wind ripples and wind speed in different weather conditions.

Both laboratory and theoretical models are constructed for a specific, rather narrow range of conditions, such as the stage of development, fetch length, the influence of dominant waves and near-surface currents, etc. The question of how often the conditions for which the model is constructed during field measurements are realized, and also to what extent the level of spectral density changes when the specified conditions change, remains open.

Studies of the relationship between the energy of short-period waves and wind speed were carried out on the oceanographic platform of the Marine Hydrophysical Institute. Elevations of the sea surface were measured by a broadband string waveograph [35]. During the measurement period, the wind speed varied between 0 and 20 m/s, the height of dominant wind waves and swell varied from 0.3 m to 2.5 m. The mean-square amplitudes of the spectral components of sea waves $A(f_i)$ were determined using a ten-channel parallel spectrum analyzer (i—the channel number). The frequency range in which the waveform signal is analyzed is 0.65–23 Hz. The ratio of the frequency bandwidth Δf_i to the center frequency f_i is 0.27 [21].

Synchronous measurements of sea waves and wind speed were carried out with sessions lasting 0.5 min. The duration of a series of consecutive sessions was 1–2 h. During further processing, the data were averaged over 10 points, which corresponds to a period of 5 min. Statistical analysis was performed for the parameters $A(f_i)$ and U obtained in this way.

A measure of the linear statistical relationship of the two parameters is the correlation coefficient. The correlation coefficients r_i between the parameters $A(f_i)$ and wind speed are shown in Fig. 10.6. A pronounced growth trend r_i is visible with increasing frequency.

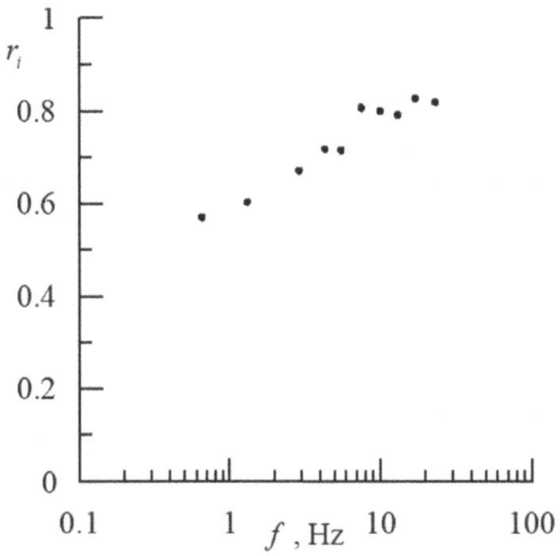

Fig. 10.6 Correlation coefficients r_i characterizing the dependence of the amplitudes of the spectral components of the waves on the wind speed

The spectrum of surface waves is a steeply decreasing function of frequency. The root-mean-square amplitudes of the spectral components $A(f_i)$ are very different, normalization is introduced to compare their behavior at different wind speeds

$$\tilde{A}(f_i) = A(f_i) / \langle A(f_i) \rangle \tag{10.20}$$

where symbol $\langle\rangle$ means the average value $A(f_i)$ at wind speeds of 8–10 m/s. The calculation results $\tilde{A}(f_i)$ are presented in Fig. 10.7.

Since it was shown above that the correlations between the spectral amplitudes of high-frequency waves and wind speed are quite high, we can construct regression dependences. Linear regressions were originally calculated

$$\tilde{A}(f_i) = a_0(f_i) + a_1(f_i)\,U \tag{10.21}$$

The coefficients a_0 and a_1 Eq. (10.21) are given in Table 10.2. The coefficients a_1 determining the response of the spectral components to fluctuations in wind speed turned out to be higher at frequencies from 5.5 to 23 Hz, where $0.58 < a_1 < 0.77$, while at frequencies $0.65 < f_i < 4.3$ Hz their values were below $0.51 < a_1 < 0.56$. This means that short gravitational waves with frequencies of 0.65–4.3 Hz are less sensitive to oscillations U than gravitational-capillary waves with frequencies of 5.5–23 Hz.

To quantify deviations from linear regression, polynomial-type quadratic regression equations were constructed

$$\tilde{A}(f_i) = b_0(f_i) + b_1(f_i)U + b_2(f_i)U^2 \tag{10.22}$$

The values of the second-order polynomial regression coefficients are also presented in Table 10.2. It can be seen that, at all frequencies, the ratios between the

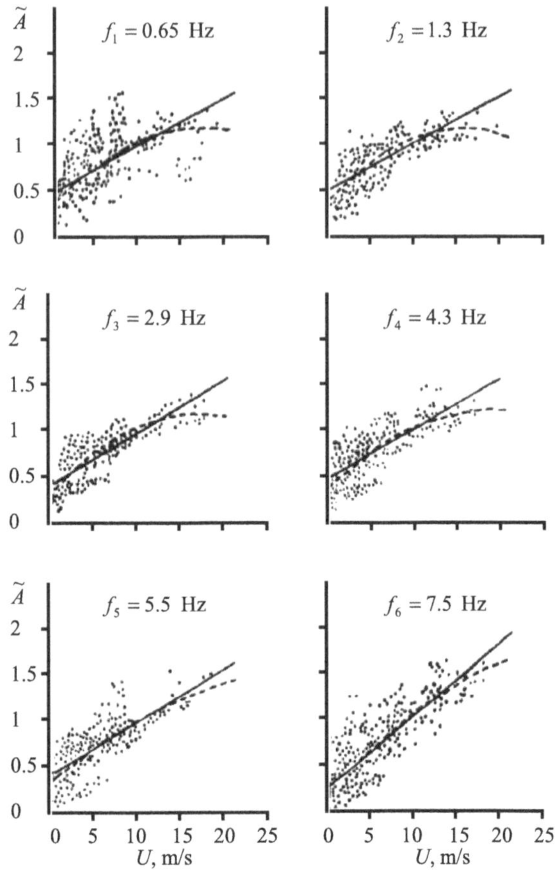

Fig. 10.7 Dependences of the normalized mean square amplitudes of the spectral components $\tilde{A}(f_i)$ of sea waves on the wind speed U. The solid and dashed lines correspond to the regression Eqs. (10.21) and (10.22)

quadratic and linear terms b_2/b_1 do not exceed 0.032, that is, the linear approximation in the form (10.21) can be considered quite satisfactory.

Let us estimate the scatter of values $A(f_i)$. The scatter of the values of the random variable x characterizes the coefficient of variation

$$K = \sqrt{\overline{x^2}} \Big/ \overline{x} \tag{10.23}$$

where $\overline{x^2}$ is variance; $\overline{x}$ is average value. The coefficients of variation K_A for $A(f_i)$ are given in Table 10.3. Since the scatter of values $A(f_i)$ changes with the change in wind speed, the range of wind speeds at which $A(f_i)$ were measured is divided into 8 intervals that correspond to points from 1 to 8 on the Beaufort scale. The values of the coefficients of variation are presented in Table 10.3. Here N indicates the number of measurements in each interval of wind speeds.

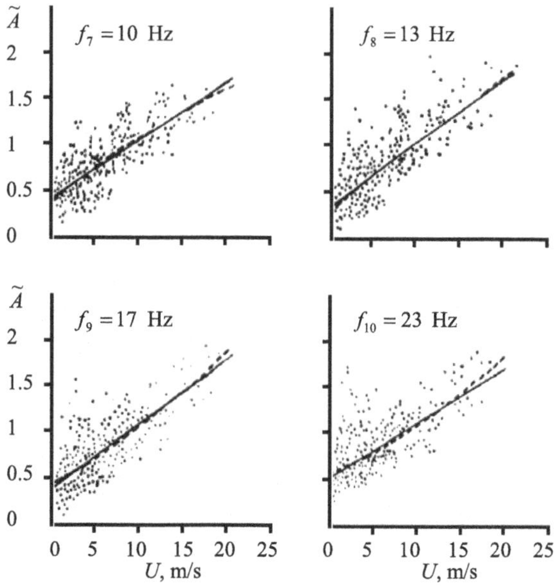

Fig. 10.7 (continued)

Table 10.2 The coefficients of the regression Eqs. (10.21) and (10.22)

f_i, Hz	Linear regression $\tilde{A} = a_0 + a_1 U$		Second-order polynomial regression $\tilde{A} = b_0 + b_1 U + b_2 U^2$			
	a_0	a_1	b_0	b_1	b_2	b_2/b_1
0.65	0.445	0.052	0.351	0.090	−0.0025	−0.028
1.3	0.482	0.051	0.357	0.102	−0.0033	−0.032
2.9	0.502	0.051	0.395	0.094	−0.0028	−0.030
4.3	0.455	0.056	0.374	0.089	−0.0021	−0.024
5.5	0.429	0.059	0.384	0.077	−0.0012	−0.016
7.5	0.276	0.077	0.224	0.099	−0.0014	−0.014
10.0	0.405	0.062	0.381	0.071	−0.0006	−0.009
13.0	0.334	0.071	0.343	0.067	0.0002	0.004
17.0	0.399	0.070	0.428	0.059	0.0007	0.012
23.0	0.519	0.058	0.546	0.048	0.0007	0.015

The values of the coefficients of variation vary from 0.04 to 0.47. There is a clear tendency toward a decrease in the coefficient of variation with increasing wind speed, which indicates greater stability of the statistical characteristics of the high-frequency components of the wave field under strong winds compared with weak and moderate winds. The physical reasons, apparently, are related to the fact that as the wind weakens, it ceases to play a dominant role in the formation of roughness on the

Table 10.3 Coefficient of variation K_A

Wind speed (m/s)	Frequencies of spectral components f_i (Hz)										N
	00.65	10.3	20.9	40.3	50.5	70.5	10	13	17	23	
0.25–1.54	00.39	00.41	00.38	0.38	0.42	00.47	0.28	00.33	00.29	00.22	368
1.55–3.34	00.31	00.29	00.23	0.24	0.25	00.41	00.29	00.41	00.36	00.31	700
3.35–5.44	00.32	00.22	00.21	00.25	00.31	00.37	00.28	00.35	00.27	00.22	647
5.45–7.94	00.33	00.17	00.17	00.21	00.27	00.33	00.23	00.29	00.23	00.20	531
7.95–10.49	00.24	00.13	00.13	00.19	00.20	00.19	00.24	00.21	00.20	00.31	308
10.50–13.84	00.17	00.07	00.08	00.11	00.10	00.17	00.14	00.17	00.15	00.16	231
13.85–17.14	00.15	00.06	00.06	00.07	00.10	00.11	00.12	00.12	00.11	00.11	100
17.15–20.75	00.13	00.04	00.04	00.05	00.06	00.06	00.07	00.07	00.06	00.06	34

sea surface, since internal waves and currents that develop various inhomogeneities, such as surface spots, areas with higher roughness etc. begin to play a commensurate role.

10.3.2 *Long Wave Effect*

There are several definitions of the concepts of "long" and "short" waves. Here, long is meant waves having a length comparable to the length of the main energy-carrying waves that form a peak in the elevation spectrum. Accordingly, short waves are waves with wave numbers and frequencies that are much larger than those of dominant waves. A characteristic feature of short waves is a short relaxation time, they quickly adapt to changes in wind speed and direction, which allows them to be used for remote determination of driving wind. Changes in the energy of short waves, as shown above, are well correlated with changes in wind speed. The energy of long waves, in addition to being dependent on wind speed, also depends on the acceleration length and duration of the wind. Let us consider how the variances of slopes created by long waves at a fixed wind speed in different situations differ.

In 1973, under the leadership of Professor C. Hasselman, a group of European and North American scientists carried out a project in the North Sea called the "Joint North Sea Wave Project" (JONSWAP) [17]. His goal was to study the generation and development of sea wind waves. One of the results of this project was the construction of an empirical frequency spectrum of the surface wave field under conditions of limited fetch. This spectrum $S(\omega, X, U)$, called the JONSWAP spectrum, depends on the wind speed U and fetch length X. The spectrum $S(\omega, X, U)$ has the form

$$S_J(\omega, X, U) = \frac{a\, g^2}{\omega^5} \exp\left(-\frac{5}{4}\left(\frac{\omega_0}{\omega}\right)^4\right) \gamma^{\Phi} \tag{10.24}$$

where ω is circular frequency; g is gravitational acceleration;

$$a = 0.076\left(\frac{W_{10}^2}{X\,g}\right)^{0.22};\ \omega_0 = 22\left(\frac{g^2}{W_{10}\,X}\right)^{0.33};\ \gamma = 3.3;$$

$$\Phi = \exp\left(-\frac{(\omega-\omega_0)}{2\,\upsilon^2\,\omega_0^2}\right);\ \upsilon = \begin{cases} 0.07 & npu \quad \omega \le \omega_0 \\ 0.09 & npu \quad \omega \le \omega_0 \end{cases};$$

where W_{10}—wind speed measured at a height of 10 m, γ^{Φ} is characteristic describing the excess in the peak region relative to the Pearson-Moskowitz spectrum [29]; υ_J is parameter determining the width of the spectral peak.

The JONSWAP spectrum is similar to the Pearson-Moskowitz spectrum, which was built for a situation where the wind wave is fully developed. The difference is that the peak of the JONSWAP spectrum is more pronounced, and it increases with increasing wavelength (or the duration of the wind). If we accept $\gamma = 1$, then the JONSWAP spectrum will coincide with the Pearson-Moskowitz spectrum. The JONSWAP spectrum was determined under the following conditions: limited acceleration, constant wind and no swell. It was widely used and was used even in situations where the conditions for the generation of surface waves did not correspond to the conditions under which it was obtained.

When constructing the spectrum $S(\omega, X, W)$, the data of wave measurements were used, obtained in situations when the wind is stable in speed and direction, and there is no swell. Such clean conditions are rarely observed in the open ocean and in coastal zones. It was noted in Danilychev et al. [9] that according to various studies performed in different regions of the World Ocean, the proportion of wind waves can range from 35 to 45%. This assessment includes developing, developed, and damped wind waves. Conditions that are close to the situation of developed wind turbulence are approximately 5%.

We pass from the spectrum of elevations of the sea surface to the spectra of its slopes. The spectra of the upwind $\Psi_u(\omega)$ and crosswind $\Psi_c(\omega)$ components of the slopes are related to the frequency-angular spectrum of elevations of the sea surface by the relations

$$\Psi_u(\omega) = k^2(\omega)\int_0^{2\pi} \chi(\omega,\varphi)\,\cos^2\varphi\,d\varphi \tag{10.25}$$

$$\Psi_c(\omega) = k^2(\omega)\int_0^{2\pi} \chi(\omega,\varphi)\,\sin^2\varphi\,d\varphi \tag{10.26}$$

where $k(\omega)$ is wave number; $\chi(\omega, \varphi)$ is frequency-angle spectrum; φ is direction of wave propagation. In deriving Eqs. (10.25) and (10.26), it was assumed that the general direction of wave propagation corresponds to the condition $\varphi = 0$. Imagine the frequency-angle spectrum $\chi(\omega, \varphi)$ in the form

$$\chi(\omega, \varphi) = S(\omega)\,\Theta(\omega, \varphi) \tag{10.27}$$

where $\Theta(\omega, \varphi)$ is angular distribution function of wave energy satisfying the normalization condition $\int_0^{2\pi\,2\pi} \Theta(\omega, \varphi)\Theta(\omega, \varphi)\, d\varphi = 1$.

We get

$$\Psi_\xi(\omega) = \Psi_u(\omega) + \Psi_c(\omega) = k^2(\omega)\, S(\omega) \tag{10.28}$$

We assume that the waves propagate through "deep" water, i.e. the influence of the bottom can be neglected. In this case, the long waves obey the dispersion equation

$$\omega^2 = g\,k \tag{10.29}$$

The transition from the spectrum of elevations to the spectrum of slopes of the sea surface in the final form is described as

$$\Psi_\xi(\omega) = \frac{\omega^4}{g^2}\, S(\omega) \tag{10.30}$$

To estimate the dependences of the variances of slopes created by long waves on the acceleration length, we introduce the parameter

$$\overline{\xi_b^2} = \int\limits_0^{\omega_b} \Psi_\xi(\omega)\, d\omega \tag{10.31}$$

where ω_b is the upper limit of the frequency range for which the variance of slopes is calculated. The results of calculating the dependence of the slope variance determined in the framework of the JONSWAP spectrum model for different accelerations are presented in Fig. 10.8.

In the case when the variance $\overline{\xi_b^2}$ is determined in the range with the upper frequency limit of $\omega_b/(2\pi) = 1$ Hz (the indicated frequency corresponds to a wavelength of 1.6 m), it rapidly grows with an increase in the acceleration length to 2000–3000 m and practically does not change at $X > 5000$ m (see Fig. 10.8). If $\omega_b/(2\pi) = 0.5$ Hz (corresponds to a wavelength of 6.3 m), then the saturation state occurs later, but in this case the values $\overline{\xi_b^2}$ are significantly, about one and a half times lower.

It is seen that at the same wind speed, the variance of the slopes created by long waves can differ several times. Accordingly, the parameters of the distribution of slopes change, and, as follows from (10.19), long waves, depending on the stage of development of the wave field, will affect the signal of the scatterometer to varying degrees.

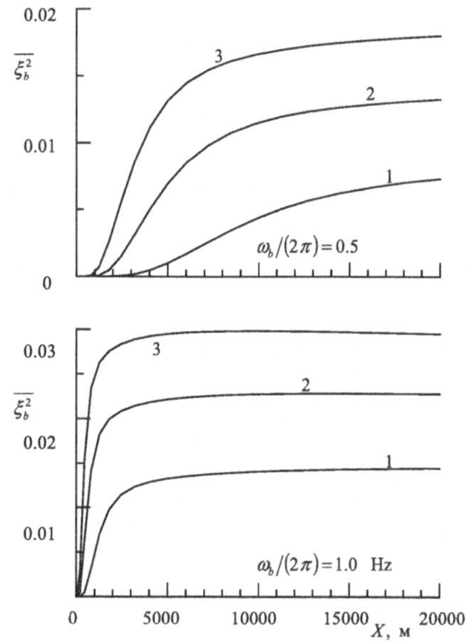

Fig. 10.8 Dependence of the variance of long wavelength slopes σ_b^2 on the length fetch X. Curves 1–3 correspond to wind speeds of 5, 10 and 15 m/s

10.4 Conclusion

Remote methods for measuring air flow over the sea are based on the sea surface roughness depending on the wind speed, which determines the power of the radio signal reflected from it. Direct measurements of the characteristics of wind ripples, such as the variance of elevations and the variance of slopes showed that the values of these parameters increase with increasing wind speed. Their connection with wind speed is statistical. In the range of 4–15 m/s this makes it possible to restore the wind field from the measured ripple characteristics with an accuracy of $\pm$1–2 m/s.

In conditions when the wind does not exceed 4–5 m/s, with a one-parameter restoration of wind by ripples, the error becomes many times higher than with contact anemometry. One of the immediate reasons for this is the fluctuation in the spectral density levels of ripples during passage through the smoothing zones (slicks) and increased roughness against the average ripple background corresponding to a given wind speed. Such oscillations are associated with the manifestation of the paranasal surface of internal waves and other strong mechanisms that are manifested during neutral stratification of the atmospheric boundary layer, when the dynamic influence of wind ceases to play a dominant role.

When conducting direct sub-satellite measurements of wind speed over the sea, an additional fatal error appears, the existence of which is associated with spatio-temporal inhomogeneities of the wind field. Even in the case of a very stable wind, the relative level of turbulent fluctuations (turbulence intensity) reaches 10%, which leads to an ambiguity of about 1 m/s at a wind speed of 10 m/s. If internal waves propagate in the atmosphere, the period of which is usually 10–20 min, and their direction may differ from the direction of the local wind, then fluctuations in wind speed can reach, for example, 2 m/s at an average wind speed of about 5–7 m/s; this leads to a corresponding scatter between the values of wind speed and direction obtained from remote sensing data, compared with direct observations on buoys.

This circumstance limits the real accuracy of the sub-satellite calibration (and, therefore, the accuracy of reconstructing the wind field) to the level of the mean square error to 0.5–1 m/s at high wind speeds and several times greater error at low wind speeds, $U < 4$ m/s, since in this case the wind is often unstable due to an increase in the relative role of convection and wave processes in the atmosphere.

The results obtained show the existence of physical limitations in the accuracy of recovery of wind speed by means of remote sensing, which are especially noticeable in weak and temperate winds. At the same time, altimetry and scatterometer methods for determining the parameters of the air flow over the ocean have an undeniable advantage. They can provide sufficiently high-quality information on a global scale, which until now was not available using traditional methods.

References

1. Apel JR (1994) An improved model of the ocean surface wave vector spectrum and its effects on radar backscatter. J Geophys Res 99(C8):16269–16291
2. Babanin AV, Polnikov VG (1994) On non-Gaussian wind waves. Phys Oceanogr (3):79–82
3. Bass FG, Fuks IM (1979) Wave scattering by statistically rough surface. Pergamon, 540p
4. Bréon FM, Henriot N (2006) Spaceborne observations of ocean glint reflectance and modeling of wave slope distributions. J Geophys Res Oceans 111(C06005). https://doi.org/10.1029/2005jc003343
5. Brown GS, Stanley HR, Roy NA (1981) The wind speed measurements capability of spaceborne radar altimetry. IEEE J Oceanic Eng 6:59–63
6. Chelton DB (1988) WOCE/NASA altimeter algorithm workshop. U. S. WOCE Technical Rep. N. 2. U. S. Planning Office for WOCE. College Station, 70p
7. Chu X, He Y, Karaev VY (2012) Relationships between Ku-Band radar backscatter and integrated wind and wave parameters at low incidence angles. IEEE Trans Geosci Remote Sens 50(11):4599–4609. https://doi.org/10.1109/tgrs.2012.2191560
8. Cox C, Munk W (1954) Measurements of the roughness of the sea surface from photographs of the sun glitter. J Optical Soc Am 44(11):838–850
9. Danilychev MV. Kutuza BG. Nikolayev AG (2009) The use of the Kirchhoff method for practical calculations in microwave radiometry of rough sea surface. Radiotekh i Elektron J Commun Technol Electron 54(8):915–926
10. Dobson E, Monaldo F, Goldhirsh J, Wilkerson J (1987) Validation of Geosat altimeter-derived wind speeds and significant wave heights using buoy data. Johns Hopkins APL Tech Dig 8:222–233. https://doi.org/10.1029/jc092ic10p10719
11. Ebuchi N, Graber HC, Caruso MJ (2002) Evaluation of wind vectors observed by QuikSCAT/sea-winds using ocean Buoy data. J Atmos Oceanic Technol 19(12):2049–2062. https://doi.org/10.1175/1520-0426(2002)019%3c2049:EOWVOB%3e2.0.CO;2
12. Freilich MH, Dunbar RS (1999) The accuracy of the NSCAT 1 vector winds: comparisons with National Data Buoy Center buoys. J Geophys Res 104(C5):11231–11246. https://doi.org/10.1029/1998JC900091
13. Freilich MH, Vanhoff BA (2003) The relationship between winds, surface roughness, and radar backscatter at low incidence angles from TRMM precipitation radar measurements. J Atmos Oceanic Technol 20(4):549–562

14. Glazman RE, Greysukh A (1993) Satellite altimeter measurements of surface wind. J Geophys Res 98(C2):2475–2483
15. Gourrion J, Vandemark D, Bailey S, Chapron B, Gommenginger GP, Challenor PG, Srokosz MA (2002) A two-parameter wind speed algorithm for Ku-band altimeters. J Atmos Oceanic Technol 19(12):2030–2048. https://doi.org/10.1175/1520-0426(2002)019%3c2030:atpwsa
16. Grishechkin BY, Baskakov AI (2010) The potential of a space altimeter when measuring significant wave height, vol XXXVIII, Part 7A. In: ISPRS TC VII symposium—100 years ISPRS, Vienna, Austria, July 5–7, 2010, IAPRS, pp 49–52
17. Hasselmann K, Barnett TP, Bouws E, Carlson H, Cartwright DE, Enke K, Ewing JA, Gienapp H, Hasselmann DE, Kruseman P, Meerburg A, Mller P, Olbers DJ, Richter K, Sell W, Walden H (1973) Measurements of wind-wave growth and swell decay during the joint North Sea Wave Project (JONSWAP). Ergnzungsheft zur Deutsch Hydrographischen Z Reihe A (8)(12):1–95
18. Karaev VY, Kanevsky MB, Balandina GN, Cotton PD, Challenor PG, Gommenginger CP, Srokosz MA (2002) On the problem of the near ocean surface wind speed retrieval by radar altimeter: a two-parameter algorithm. Int J Remote Sens 23(16):3263–3283
19. Karaev VY, Kanevsky MB, Meshkov EM, Titov VI, Balandina GN (2008) Measurement of the variance of water surface slopes by a radar: verification of algorithms. Radiophys Quantum Electron 51(5):1–12
20. Khristoforov GN, Zapevalov AS, Smolov VE (1987) On the limiting accuracy of satellite scatterometer estimates of wind-speed over the ocean. Sov J Remote Sens 7(2):287–300
21. Khristoforov GN, Zapevalov AS, Smolov VE (1994) The relationship between the amplitudinal characteristics of the high-frequency spectral components of wind-generated waves and the wind velocity over the sea. Phys Oceanogr 5(3):221–230
22. Khristophorov GN, Zapevalov AS, Babiy MV (1992) Statistics of sea-surface slope for different wind speeds. Okeanologiya 32(3):452–459
23. Kudryavtsev V, Hauser D, Caudal G, Chapron B (2002) A semiempirical model of the normalized radar cross-section of the sea surface, 1, Background model. J Geophys Res 107. https://doi.org/10.1029/2001jc001003
24. Lavrova OY, Kostianoy AG, Lebedev SA, Mityagina VI, Ginzburg AI, Sheremet NA (2011) Complex satellite monitoring of the russian seas. Institute of Space Research of the Russian Academy of Sciences, Moscow, 480p
25. Liu Y, Su M-Y, Yan X-H, Liu WT (2000) The mean-square slope of ocean surface waves and its effects on radar backscatter. J Atmos Oceanic Technol 17:1092–1105
26. Longuet-Higgins MS (1963) The effect of non-linearities on statistical distribution in the theory of sea waves. J Fluid Mech 17(3):459–480
27. Moore RK, Pierson WJ (1966) Measuring sea state and estimating surface winds from a polar orbiting satellite. In: Proceedings of international symposium on electromagnetic sensing of the earth from satellites. Miami Beach, FL. November 22–24, pp R1–R28
28. Moore RK, Claassen JP, Cook AC, Fayman DL, Holtzman JC, Sobti A, Ulaby FT, Young JD, Hatcher NM, Spencer WJ (1974) Simultaneous active and passive microwave response of the earth—the Skylab radar experiment. In: Proceedings of 9th international symposium on remote sensing of environment. University of Michigan, Ann. Arbor, MI, pp. 189–217
29. Pierson WI, Moskovitz L (1964) A prosed spectral form for fully developed wind seas based on the similarity method of S.A. Kitaigorodsii. J Geophys Res 69(24):5181–5190
30. Queffeulou P (2004) Long-term validation of wave height measurements from altimeters. Mar Geodesy 27:495–510
31. Quilfen Y, Chapron B, Vandemark D (2001) The ERS scatterometer wind measurement accuracy: evidence of seasonal and regional biases. J Atmosp Oceanic Technol 18(10):1684–1697. https://doi.org/10.1175/1520-0426018%3c1684:TESWMA%3e2.0.CO;2
32. Valenzuela G (1978) Theories for the interaction of electromagnetic and ocean waves—a review. Bound Layer Meteorol 13(1–4):61–85
33. Walsh EJ (1979) Extraction of ocean wave height and dominant wavelength from GEOS 3 altimeter data. J Geophys Res 84(B8):4003–4010. https://doi.org/10.1029/JB084iB08p04003

34. Wilson JJW, Anderson C, Baker MA, Bonekamp H, Saldaca JF (2010) Radiometric calibration of the advanced wind scatterometer radar ASCAT carried onboard the METOP-a satellite. IEEE Trans Geosci Remote Sens 48(8):3236–3255. https://doi.org/10.1109/TGRS.2010.2045763
35. Zapevalov AS (2009) Bragg scattering of centimeter electromagnetic radiation from the sea surface: the effect of waves longer than Bragg components. Izv Atmos Oceanic Phys 45(2):253–261
36. Zapevalov AS (2002) Statistical characteristics of the moduli of slopes of the sea surface. Phys Oceanogr 12(1):24–31
37. Zapevalov AS (2008) Statistical models of the sea surface in problems of acoustic and electromagnetic radiation scattering. Manuscript to claim the academic degree of doctor of physico-mathematical sciences on the speciality 04.00.22—geophysics. Marine Hydrophysical Institute of the National Academy of Sciences of Ukraine, Sebastopol

The manufacturer's authorised representative in the EU is Springer Nature Customer Service Centre GmbH, Europaplatz 3, 69115 Heidelberg, Germany. If you have any concerns regarding our products, please contact ProductSafety@springernature.com

Printed and bound by CPI Group (UK) Ltd, Croydon, CR0 4YY
30/09/2026
02225691-0001